Andreas Wolf Ralf Steinmann Henrik Schunk

Grippers in Motion

The Fascination of Automated Handling Tasks

Andreas Wolf Ralf Steinmann Henrik Schunk

Grippers in Motion

The Fascination of Automated Handling Tasks

Dr. Andreas Wolf
robomotion GmbH
Industriestraße 25
70565 Stuttgart
Germany

Ralf Steinmann
Schunk GmbH & Co. KG
Spann- und Greiftechnik
Bahnhofsstraße 106-134
74348 Lauffen/Neckar
Germany

Henrik Schunk
Schunk, Inc.
211 Kitty Hawk Drive
Morrisville, NC 27560
USA

Library of Congress Control Number: 2005925774
ISBN 3-540-25657-1

Springer is a part of Springer Science+Business Media

springeronline.com

Printed in Italy

Cover design: reform design Stuttgart
Typesetting by the author

Printed on acid-free paper 89/2166 – 5 4 3 2 1 0

Foreword by Prof. Joseph F. Engelberger

In America you can often hear of people who "come to grips" with a problem. That is, they are "taking hold" of the situation.

And, it is "gripping" which is so critical to my discipline, robotics.

Preprogrammed automatons have been around since the fourteenth century. They are clever and entertaining, but they cannot interact with the workaday world. For example, at the 1938 New York World's Fair one of the automaton hits was "Electro" and his dog, "Sparky". Electro stumbled around, waved his arms and repeated recorded messages while Sparky hopped around and barked.

Note that neither interacted with the real world. That remained for post World War II technology and the advent of the industrial robot.

The point of the robot was that a multi-axis manipulator could be programmed to do a variety of manufacturing tasks.

Yes, the robot arm's contoller could memorize a number of complex manipulations. Still, it had to grasp work pieces, not just wave its arm.

Enter the end effector! And, enter **Grippers in Motion**. This book is magnificently illustrated. Therein you find grippers that can cope with a range of work pieces, say in an assembly task. There is even recognition of the human hand as an ultimate standard. Multiple fingered end-effectors with tactile sensing can communicate with a robot's contoller to handle disoriented work pieces.

All in all an automation designer is probably going to find precisely the gripper to meet his proposed application. Should he not at least be inspired by the search through **Grippers in Motion**, then he is rather a dull engineer.

Joseph Engelberger
Newtown Connecticut, USA, in January 2005

Foreword by Heinz-Dieter Schunk

It is fascinating how easily and sensitively the human hand as the "tool of tools" (Aristotle) virtually gets to grips with its environment as the "extension of the human brain" (Kant).

In the 15th century Goetz von Berlichingen, the "Knight with the Iron Hand", instructed creative craftsmen to integrate the complicated technique of finger movements by inventing a cable pull mechanism. Despite these early efforts the worldwide development of grippers as a kind of artificial or industrial hand will probably never be able to reach the perfection and overall functionality of the "human hand".

The latter remains the perfect model for a technology which as an end-effector in automation today is still a constant challenge to engineering. In their early stages, grippers had to be adapted to robot capacities in terms of optimum weight while now we are at a point where standardized components are set in motion as reliable "hands" in automated production.

Today's grippers have developed a sense of touch with the help of sensor technology and are even able to see with the support of image processing systems. The industrial hand together with related technological developments is gradually on the move towards copying the perfectly multi-functional human model.

The publication "Grippers in Motion" can only be a snapshot of state-of-the-art technology and illustrates the development of gripping technology to the interested reader. At the same time, it shows how great the variety and opportunities of gripper applications are today.

I would like to extend my thanks to all who have contributed to this book and wish the reader many interesting incitations.

Heinz-Dieter Schunk
Lauffen am Neckar, Germany, in May 2004

Preface

Fascinating automation technology was the motivation for writing this book. Every day we are faced with new developments in automation and great opportunities for its applications. Intelligent gripping systems and service robots are the early signs of a new, more flexible automation technology, which is capable to auto-adapt to changing environments. This publication aims to allow a close look at the ambient conditions under which this technology is used. The great variety of grippers and robot components about 30 years after the first industrial applications speaks for itself. Applications in the pharmaceutical, food processing industry, and in agricultural production, are not yet standard but do offer a growing market for automated solutions in the near future.

This publication can only cover a section of worldwide gripper and robot technology. With illustrating graphics and tables explaining the details we intend to catch the reader`s interest. Further literature and references to it are included for more detailed information. The area of conflict between the movements of grippers and those of the human arm is used in order to explain the subject to the reader with the help of analog examples from daily life. At the same time, the complexity of supposedly simple handling processes is clearly demonstrated, focusing on structuring the features of the gripping and the moving task. This structure is supported by ideograms on each page for easy guidance.

We would like to thank all who contributed to this publication. Our very special thanks go to Mr. Heinz-Dieter Schunk for encouraging us to write this book. We are very grateful to Mr. Röck and Mr. Altmann, who were of great help despite their tight schedules. We thank Mr. Müller for providing and generating graphic material. Prof. Dr.-Ing. Rolf Dieter Schraft supported us together with the Fraunhofer IPA Library.

For the English translation Mrs. Tanja Schick took care together with Mrs. Katherine Bayer and Mr. Milton Guerry form the SCHUNK Intec USA. Furthermore we want to thank everybody who gave his input to the correction work especially we want to name Mr. Frank Gaiser from SCHUNK.

Very special thanks go to Mr. Steffen Mayer and Mr. Jan Binder of robomotion GmbH, and our families who created the space for this project, which otherwise would not have been completed in due course. Last but not least we thank the whole team of "reform design Stuttgart GbR", especially Ms. Christiane Schulz as project leader and Mr. Christian Kellner, who both have been responsible for a miracle in the last few weeks of the project.

For all this work we are very grateful and hope that our enthusiasm for grippers will take as strong a grip on the readers.
For any suggestions or improvements please email us to the following address:

book@robomotion.de

The above address may also be used for collecting further applications. We would like our book to generate new ideas for handling technology.

Dr. Andreas Wolf, Ralf Steinmann and Henrik Schunk
Stuttgart, Germany, in March 2005

Table of Contents

Process Description

Terminology

Definition of Subject

1 Handling: The Underrated Process

1 Handling: The Underrated Process

Human beings can hardly be imagined without gripping or moving things in their environment. By handling objects we learn about our world and judge the objects by their volume and weight. During a lifetime this experience leads to more and more refined and efficient handling. Methods of gripping and moving have continuously improved and come so natural to us that we take handling for granted.

Most people have been able to gain vast practical experience and are well versed in gripping and moving objects. Special skills for particular movements can be trained and optimized. Professional "manual workers" are usually required to have such special skills distinguishing them from untrained amateurs by high precision, fast processing and good coordination. In this respect the handling process in connection with the appropriate tools can be considered an art. The process of human beings developing gross and fine motor skills is most evident in children. Extreme boosts of kinetic ability can be measured in terms of quality as well as in terms of quantity.

During the first months up to the age of four a great variety of movement patterns are trained for adoption, while the motor skills development between the age of four and seven focusses on perfectioning the movements. The quantitative increase in efficiency can be determined by timing a 40m run, which a four-year-old boy covers within an average 16.6 seconds. A seven-year-old takes an average 9.8 seconds for the same distance, which equals an efficiency increase of 169 percent. Differences within the same age groups in terms of movement quality are obvious: A four-year-old boy is able to catch a ball as long as the ball is played chest-high into his hands, while a seven-year-old is capable of catching the ball played waist-high or head-high and able to combine the movements of catching and throwing the ball. (Source: Kurt Meinel)

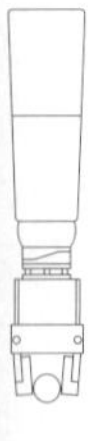

Characterization	Class (age)		Phase of ...
	♂	♀	
Newborn	0.1 - 0.2		non-directional mass movements
Baby	0.4 - 1.0		adopting first coordinated movements
Toddler	1.1 - 3.0		adopting manifold movement patterns
Early childhood	3.1 - 6/7		perfectioning of manifold movement patterns and adoption of first combined movements
Medium childhood	7.1 - 9/10		fast progress in motor learning aptitude
Late childhood	10/11 - 11/12	10/11 - 12/13	best motor learning aptitude
Early youth (pubescence)	11/12 - 13/14	12/13 - 14.5	restructuring kinetic skills and proficiencies
Late youth (adolescence)	13/14 - 17/18	14.6-18/19	developing gender-specific differentiation, progressive individualization and increasing stabilization
Early adulthood	18/20 - 30		relative maintenance of learning aptitude and kinetic performance
Medium adulthood	30 - 45/50		gradually declining kinetic performance
Later adulthood	45/50 - 60/70		considerably declining kinetic performance
Later adulthood	from 60/70		distinctly declining kinetic performance

Table 1.1 Phases of human motor skills

Table 1.1 details human motor skills changing throughout a lifetime.

For the technical recording of kinetic processes, a distance-time diagram is used which depicts illustration of various movement features. This method can also be used for determining the fine motor skills necessary for gripping workpieces.

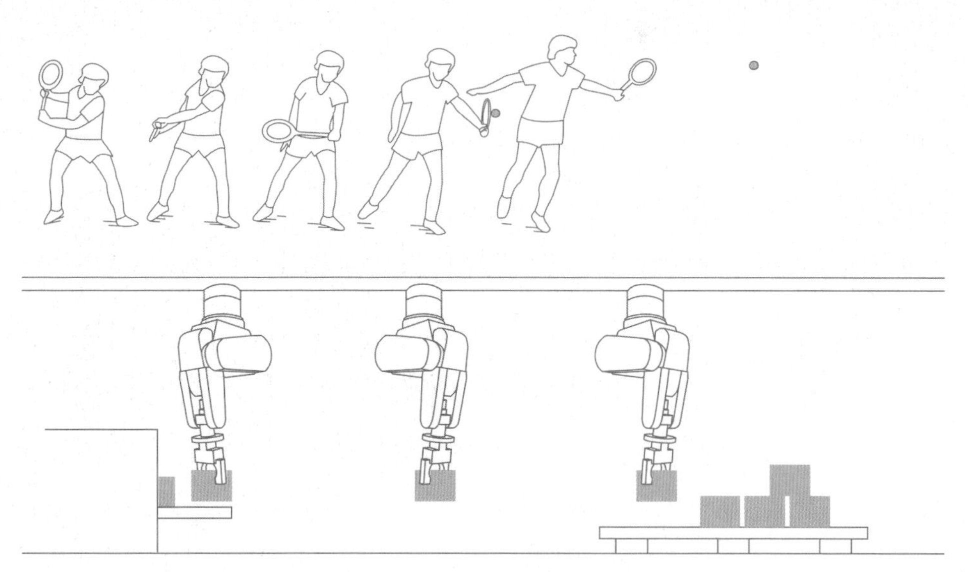

Figure 1.1 Motion sequence human/robot

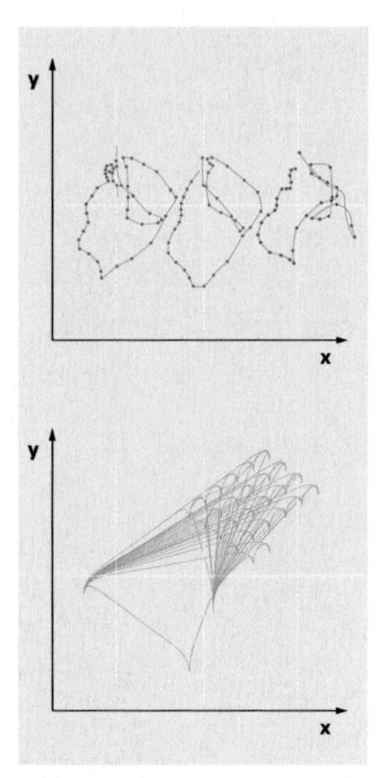

Comparison: Distance-time diagram "human" (above) and "robot" (below)

Up to now technical systems such as machines have been constructed, built and further developed for a particular movement pattern. This specialization always led to the problem of limited operating flexibility. Therefore, manufacturers aim to integrate electronic controls, new materials and sensors to increase machinery flexibility although they are faced with the problem of meeting the requirements for all applications. As a result, complex solutions comprise various components, which again are special high-tech elements, in order to meet the requirements of each application as best as possible.

Industrial manufacturing requires an efficient performance comparable to that of of top-ranking athletes. In terms of high productivity, robots are constantly improved for process reliability at reasonable prices.

Movement applications and components are frequently expected to perform around the clock which exceeds a professional athlete's set task by far. A gripper required to move workpieces at a cycle time of 15 pieces per minute performs more than 12 million opening and closing strokes per year under constant operation. Robots performing over 100 and more handling tasks per minute are quite common today.

In the 60s workpiece handling as a "a constant source of waste" was already put at the center of research by Prof. Dolezalek in Stuttgart, Germany. In terms of rationalization the focus was set on avoiding pre- and post-operating times during machine tool operations. Machine operating staff was to be decoupled from machine time by mechanical or automated systems while handling time was to be reduced to a minimum.

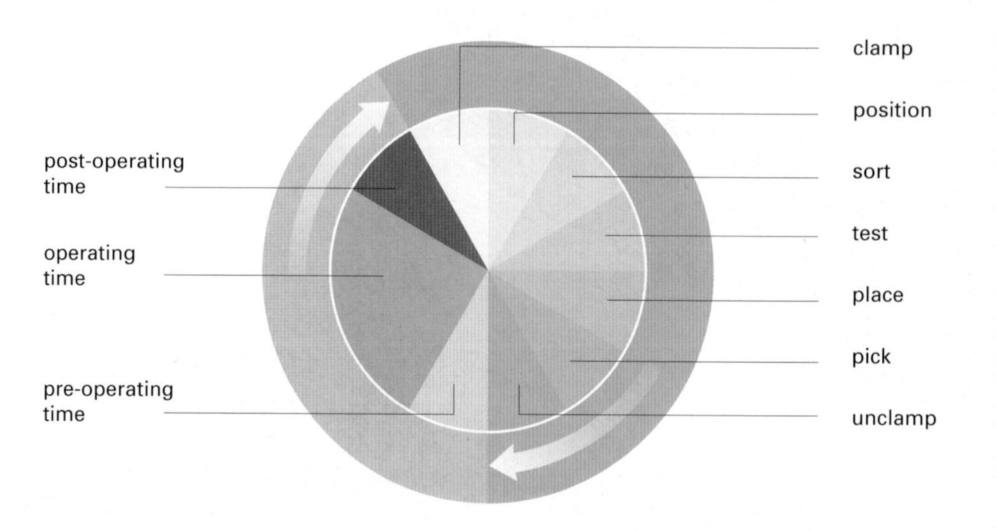

Figure 1.2 Machine time/handling time

1.1 The Handling Process

The process of handling component parts or workpieces in production is often underrated as technically simple or even trivial.
From the production point of view it is obvious that the workpiece itself does not increase in value during the handling process.
As far as technical solutions are concerned, handling is secondary to the manufacturing process. The time necessary for production is separated into machine time and handling time (see figure 1.2).

Machine time is the period of time during which a machine is operating, i. e. making changes to the workpiece itself. Machine time can be further separated into pre-operating time, operating time, and post-operating time. Pre- and post-operating time include all necessary operations before and after operating time, such as supplying a tool or coolant. These intervals have been reduced to a minimum by high traverse rates and appropriate control technology over the past few years.

Handling time or auxiliary process time can be separated into single steps from setting up a workpiece to testing it. Production planning aims at synchronizing handling time and machine time in order to prevent time-consuming handling processes from taking up valuable machine time; or at least to keep handling time at a minimum and to move as many workpieces as possible per time unit.
Machine time and handling time have to be coordinated: Machinery idling during workpiece handling is generally not acceptable, just as fast robots waiting for machinery do not make sense.

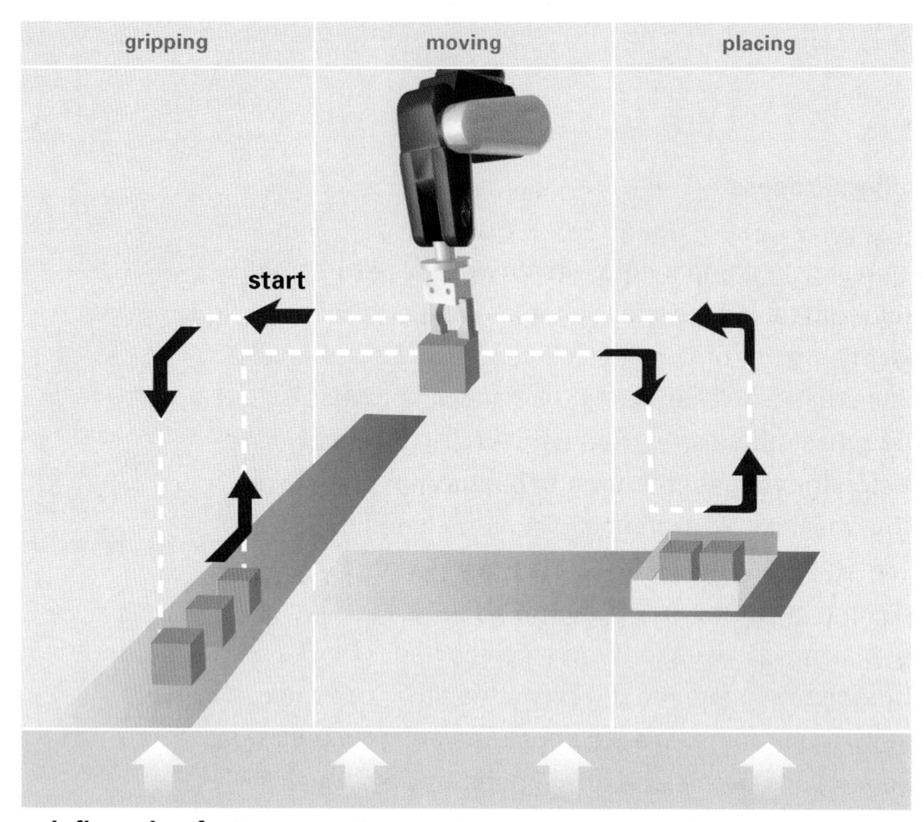

Influencing factors
- ambient conditions: type, temperature
- workpieces: orientation, quantity, position, size, type

Figure 1.3 Phases of a handling process and its ambient conditions

The handling process can be basically characterized by counting the workpieces moved per unit of time. This characteristic, however, does not specify the amount of technical requirements for obtaining a desired cycle time. Complex workpieces and multiple ambient conditions can create different handling tasks to such an extent that a simple task of moving a workpiece from point A to B can become an extremely complex process. Human beings are naturally equipped with an enormously flexible "gripping technique", efficient "sensors" and highly complex "data processing" and, therefore, tend to underrate such tasks.

From practical experience in automation projects we know that unexpected technical and economic problems tend to occur especially when the handling process and all its parameters are not sufficiently analyzed and evaluated at an early stage.

This book intends to show how the automation of handling tasks can be mastered on the basis of technical know-how and appropriate component parts. Know-how of the influencing factors of a handling process can be divided into gripping task features and moving task features.

The gripping task is mainly determined by the workpiece, its features, and the workpiece status, i .e. how it is situated.

The moving task is influenced by the features of the workpiece/gripper combination. In addition, other criteria closely related to the movement have to be met.

The gripping task and the moving task are dependent on each other. If a workpiece/gripper combination weighs too much then a particular moving task can hardly be managed or not be coped with at all.

Both gripping and moving task are subject to the same ambient conditions of the production process, such as high/low temperature or harsh environments.

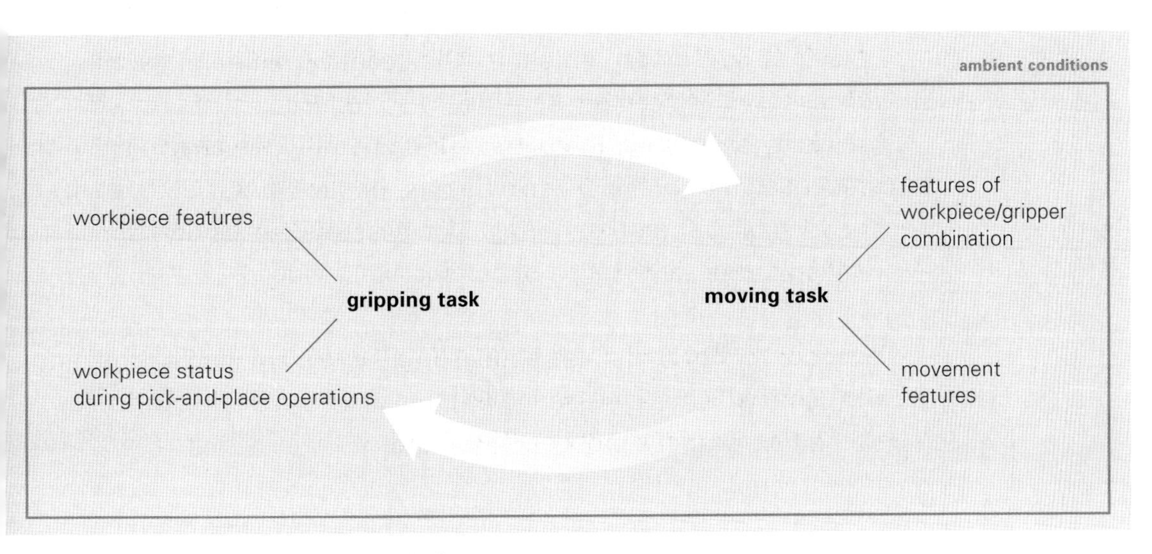

Figure 1.4 The handling process (structural overview)

1.2 Handling Terminology

The following definitions are given to introduce the reader to the technical terms of handling and are useful for understanding the following chapters. The order of terms approximately follows the production process.

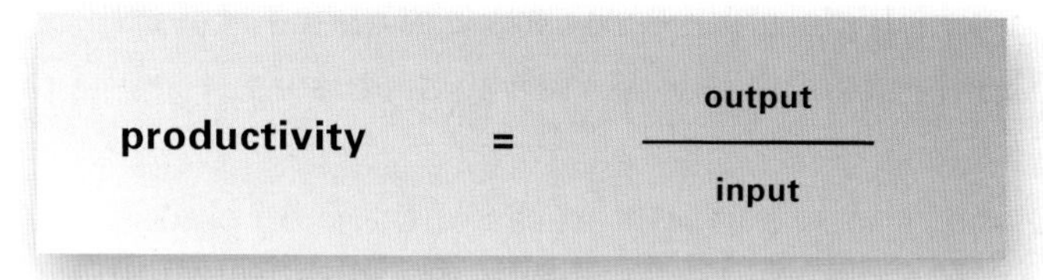

Input in relation to a given output is defined as ***productivity***.

Example:

$$\textbf{work productivity} = \frac{\textbf{turnover per year}}{\textbf{total workforce}}$$

Flexibility

Versatility and adaption flexibility are to be distinguished. Versatility is defined as a measurement for the quantity of different work processes which can be managed by a production system. Versatility expresses, therefore, the probability of managing any production task within a defined range of such tasks. Comparing the versatility of different production systems makes only sense in relation to the same range of parts to be produced.

Adaption flexibility defines a measurement for a period of time and the costs which accrue during the transfer from one work process to the next within a production system.

Workpiece

The term is taken from the field of mechanical processing. As a rule the workpiece undergoes a physical manipulation during processing. When moving workpieces the focus is not on processing itself but on changing the position and orientation of the workpiece. In this context the workpiece is often called the handling object, synonyms are also work object, product, processing part, or component part. In the following, the term workpiece will be used regardless of whether the component part is processed, moved or oriented.

Workpiece: Flexible drive shaft

Workpiece compound

A distinction must be made between a single workpiece, such as one that is without other workpieces nearby, and between a compound. The compound is defined when the workpieces are adjoined on one level or when they are stacked on top of each other.

Tool

Tools are used for processing workpieces. A general distinction during handling is made between handling the tool and handling the workpiece.

Examples of workpieces

Setup

The term is used in connection with the availability of workpieces. A workpiece is made ready when it is able to be gripped.

Handling

Handling means creating, defined changing or temporarily maintaining a pre-set alignment of geometrically defined bodies in a system of coordinates. Further parameters such as time, quantity, and path can be pre-set (source: VDI Verein Deutscher Ingenieure Guideline 2860).

Handling is a subfunction of the materials flow and categorized on the same level as conveying or storing.

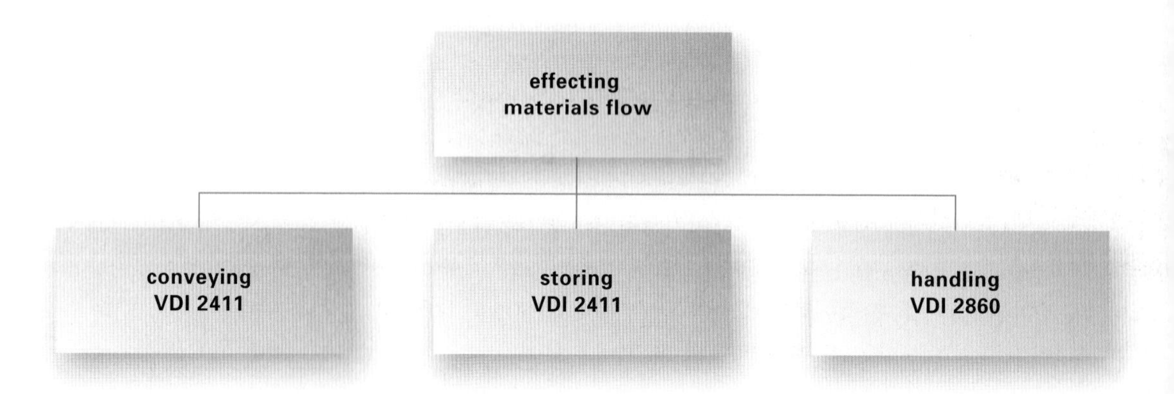

Handling is divided into the following single steps:

- Store
- Change Quantities
- Move
- Secure
- Control

In table 1.2 these single steps of handling are further divided into basic functions and complex functions. All functions are related to the workpiece or the workpiece compound.

	Store	**Change Quantities**	**Move**	**Secure**	**Control**
Basic functions		• divide • unite	• rotate • shift	• hold • release	• test
Complex functions	• stored	• separate • allocate • branch • join • sort	• swivel • orient • position • arrange • guide	• clamp • unclamp	• measure • count
Realization	• conveyor belt • pallet • magazine	• isolation device • allocator • switch	• rotary device • linear axis • industrial robot	• parallel jaw gripper • integrator • spanner	• testing device • measuring device • sensor

Table 1.2 Partial functions of handling (source: VDI Guideline 2660)

While the partial functions Store and Change Quantities mainly occur during workpiece setup, the functions Move Secure are encountered when changing the workpiece`s order status.
The Control function is an additional function which is regularly associated with handling. Just as a humans would examine the workpieces when picking them up, the workpieces in an automated process have to be monitored, too.

For the Move and Secure functions it is essential how the workpieces are prepared and presented. Various technical options are shown in the bottom line of table 1.2. The Save function for workpieces can be realized by conveyor belt, pallet, or magazine. These saving options differ according to saving volume, saving costs and workpiece order status. The function Change Quantities may have to be integrated as an interim process to be able to grip the workpiece, e. g. using vibrating devices or allocators for isolating the workpiece.

In order to move a workpiece from point A to B both partial functions **Secure and Move** are required. The technical tools and components for these functions are the subject of this publication. Mechnical grippers are appropriate for securing the workpieces. Gripping and clamping devices are used in relation to the machine operation. Components for the moving process range from rotating or swiveling devices to linear axes to industrial robots.

1 Handling device / moving device
Moving devices are technical devices which enable a workpiece to be moved. This includes the simplest movements such as pure rotations or linear movements of the workpiece as well as complex movements by industrial robots.

2 Adaption / flange adapter
In order to combine the gripper module with a handling device (moving device), for example an industrial robot, adapters are needed to connect the interfaces of both systems.

3 Gripper module
Gripper module is defined as the gripping component which combines drive and kinematics. Gripper kinematics are defined as the part of the gripper module which transforms the movement of an actuator (drive) into a movement of the gripping fingers or the force-transfer elements. The gripper drive is defined as the part of a gripper module which transforms the input power into a rotary or translatory movement.

4 Workpiece
Workpiece is defined as the object that is to be gripped. The workpiece itself is not changed or manipulated by the handling.

5 Gripper finger / force transfer element
Gripper finger or force transfer element defines the part of a gripper which induces the appropriate force on the workpiece in order to retain it against the process force.

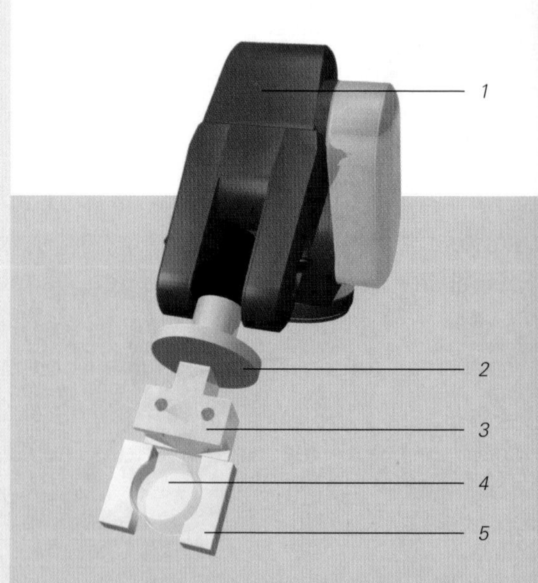

Figure 1.5 Definition of a gripper system

Gripper System

The term gripper system stands for a complex gripper which may consist of several sub-systems. The force transfer elements in a simple finger form can be configured as a complete payload system or end-effector capable of gripping one or several workpieces.

In case of differing workpiece sizes requiring different positioning of the end-effectors, drive systems (actuators) are frequently necessary to put the gripper jaws into position. Complex gripper systems require special control systems which cannot always be integrated into the gripper itself. Sensor systems provide these control systems with the necessary information for actuating the end-effectors. Appropriate safety systems are another demand on gripper technology.

As a result, completely independent gripper systems can be created with a considerable impact on the functionality of the overall production process.

The gripper system consists of several subsystems, which are not limited to the gripping process, such as:

- *payload systems*
- *end-effectors*
- *drive systems*
- *control systems*
- *sensor systems*
- *security systems*

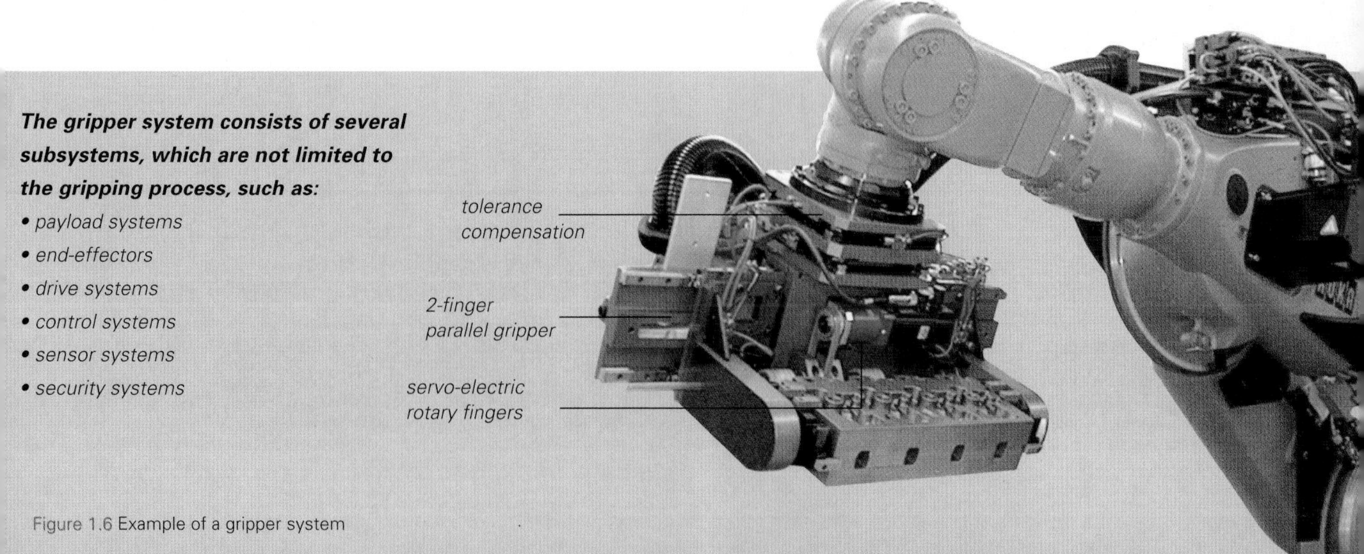

Figure 1.6 Example of a gripper system

1.3 What Are The Main Points of This Book?

The following chapters focus on the realization of handling technology tasks. Keynote is the process of integrating a workpiece into a moving device and put it into a new position or orientation.

In order to illustrate the subject in a sensible selection we draw the analogy to human object handling. In line with this analogy we concentrate on gripping techniques which follow mechanical principles (force-lock and/or form-lock). Vacuum grippers and other gripper types are included but not covered in detail. However, workpiece movement with moving axes and robot technology from the gripper finger to the six-axis robot arm are thoroughly described.

Chapter 1 explains terms and fundamentals of the subject.

Chapter 2 gives an insight into the history of automation technology and robot development over the past 30 years, highlighting the milestones without any claim for completeness. An outlook on future developments awaiting us over the next few years, supported by statistics and statements by the German federation of the engineering industries VDMA (Verband Deutscher Maschinen und Anlagenbauer) is included

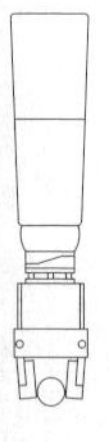

Chapter 3 explains the technical requirements for gripping, starting with the workpiece as the reason for handling. The contraints of gripping and the workpiece setup are discussed as well.

In **Chapter 4** workpiece movement as part of the gripping process is discussed. A basic knowledge of robotics that is necessary for automated movements is explained from realizing simple rotary and linear movements to multi-kinematics and robots.

Chapter 5 includes applications as provided to the authors by producers or integrators illustrating the surprising variety of automation solutions for the most diverse industries. The applications are an indicator for the practicability of automation technology and its prospects for the future.

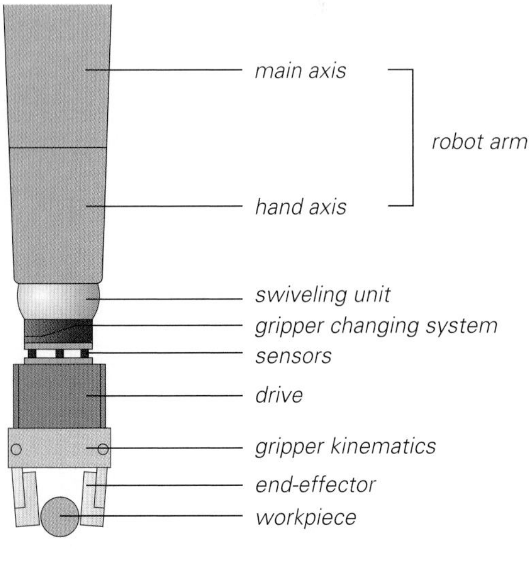

main axis
hand axis
robot arm
swiveling unit
gripper changing system
sensors
drive
gripper kinematics
end-effector
workpiece

The gripper arm and its components – an illustrated guide through this book.

Mechanical Engineering and Automation Components

Development of A New Industry

History and State-of-the-art Technology

Industrial Robot Statistics

2 Evolution or Revolution?

2 Evolution or Revolution?

In the context of new technological developments the term "revolution" is frequently applied. In fact, some of the achievements in automation technology within the past 30 years can definitely be called revolutionary. Overall statistics, however, show a rather moderate pace of development. How much of it is evolution, and how much is revolution?

The upheaval of information technology in the industrial automation arena is revolutionary. Information technology has boosted industrial automation as performance intensity was multiplied by control technology and sensor data processing. Other technological impacts are responsible for a higher concentration of output and function at decreasing costs – starting from compact actuation to lightweight kinematic structures. The overall efforts result in the output of advanced high-tech systems which are bound to be successful. The continously increasing number of robot systems clearly proves this trend.

The overall increase cannot be considered "revolutionary" when looking at robot production output figures which have been statistically recorded for more than 20 years. Despite initially low output during the mid-70s, great expectations in robotics led to the selective perception of a technical revolution. Sales not meeting exaggerated forecasts had a sobering effect. Robot applications were limited by an initially unfavorable price-performance ratio and a lack of planning capacity as an automated workshop required considerable planning efforts while trained staff and planning tools were not yet available.

UNESCO annually publishes an international survey on robot statistical development. Other institutes and institutions record robot piece numbers including detailed information on payload or number of axes. For automation technology components, such as sensors or grippers, such detailed piece numbers are difficult to determine.

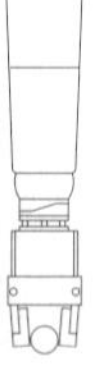

2.1 Changes in the Field of Automation

The general index by the German Federation of the Engineering Industries VDMA (Verband Deutscher Maschinen- und Anlagenbauer) presents an overview of the automation industry as a whole. This index reveals sales trends in Robotics and Automation compared to traditional Mechanical Engineering (figure 2.1).

From 1994 onwards the curve in the chart indicates a clear growth rate in automation. The main reasons for this are higher efficiency and a widespread distribution of computer technology for memory program controls such as image processing controllers or robot controllers. In the years 2002 and 2003 the growth of turnover in automation clearly exceeded that of Mechanical Engineering.

Today`s mass markets with high quantities and short production cycles, e .g. in telecommunications and computer technology, are coupled to the investment goods sector through information technology. In this respect, automation technology could be considered a technological revolution from the mid-90s onwards.

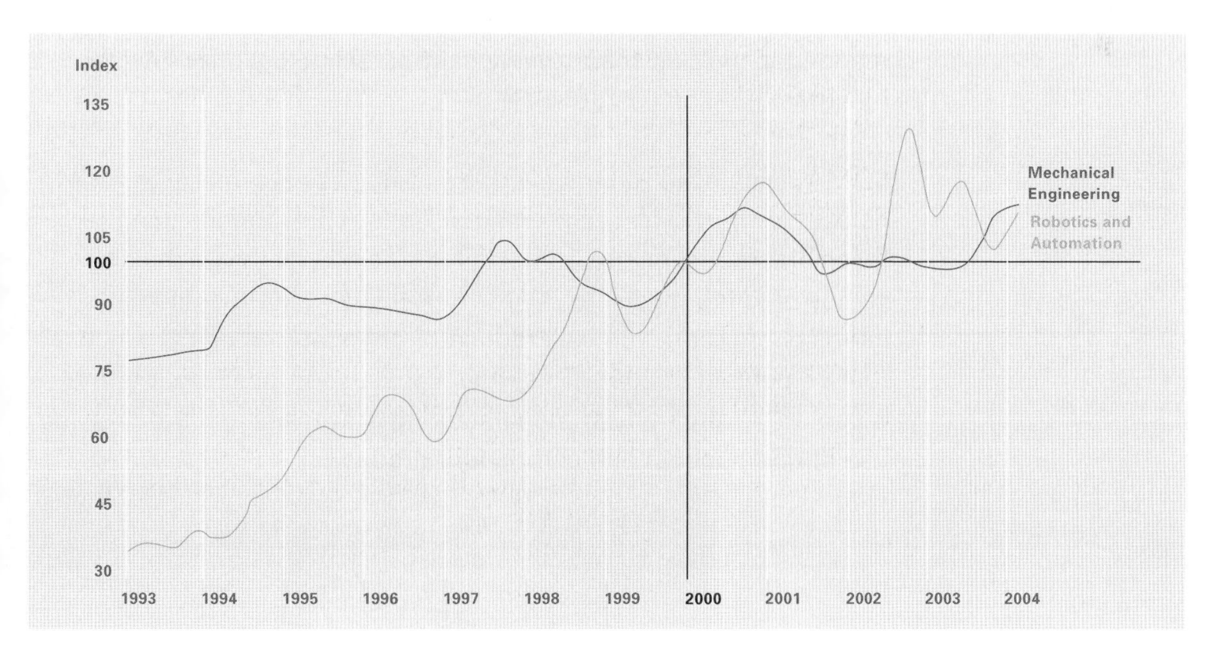

Figure 2.1 Trading volume – price and seasonally adjusted, turnover 2000 = index 100 (source: VDMA)

We will now take a close look at the evolution of grippers and robots which largely contributed to this revolution.

It is fascinating to see how parallel to progress in micro-electronics, actuator technology and mechanics – altogether called mechatronics – more and more efficient components and robots have emerged. Integrating these high-quality component parts in machine- and plant technology is standard in Mechanical Engineering today. Single components of automation technology, such as controls, servo-drives, and gripper technology, are the current basis for modern production machinery.

Mechanical Engineering in Germany has a tradition of two centuries but it is only since 1983 that the VDMA has regularly recorded operational ratios. In the past 20 years a continuous trend to reduce the average processing time from 33 days to 20 days per manufactured machine has emerged. In addition, conventional machinery has been increasingly replaced by flexible CNC machines. The share of CNC machines was at a mere 12.6 percent of the overall manufacturing machine capacity in 1983, while in 2001 it accounted for 49 percent. This is a reaction to constantly declining lot sizes in manufacturing and the connected short production runs. Automation technology, respectively control technology, has invaded manufacturing on a large scale.

VDMA ratios have to be looked at in detail as each machine technology has particular requirements. An indicator for changing machine production structures is the intensity with which companies produce machines. Within three years, the manufacturing intensity of production was reduced from about 49 percent in 1998 to 41.6 percent in 2001. This reduction can be traced back to the outsourcing of particular manufacturing processes and the growing use of components and component groups.

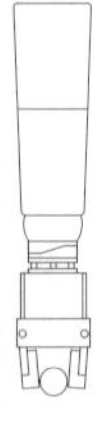

The trend illustrates a daily scenario in the automotive industry: highly innovative and complex products cannot be produced at a high value added ratio. In order to satisfy customer requirements for high-quality, constant improvement, and good price-performance ratio, it is essential to purchase structural components from specialized producers. The VDMA survey comments on the declining manufacturing intensity of production as follows:

"If products are in strong demand on the domestic market and capacities to meet this demand are not sufficient, they must be purchased on foreign markets. All enterprises which have established competent cooperation partners or suppliers in the past few years, can cope with an increase in demand. By concentrating on key competences, the prospects of playing a leading role in international competition in terms of quality, velocity, and prices, are good. Customers will naturally continue to appreciate delivery times and reliability as well as high-quality products. Short time-to-market requires enterprises to have the necessary parts ready for assembly. It is of secondary interest if these parts are manufactured in-house or purchased."

In manufacturing today, machine technology is expected to be much more flexible than it was 50 years ago. Product life cycles, i. e. the period of time during which a product is developed and marketed, is counted in months for some products of the consumer goods industry, e. g. mobile phones.

Timing its market entrance has become essential for a product`s economic success. In case of a late market entrance a product may not be able to cover its development expenses.

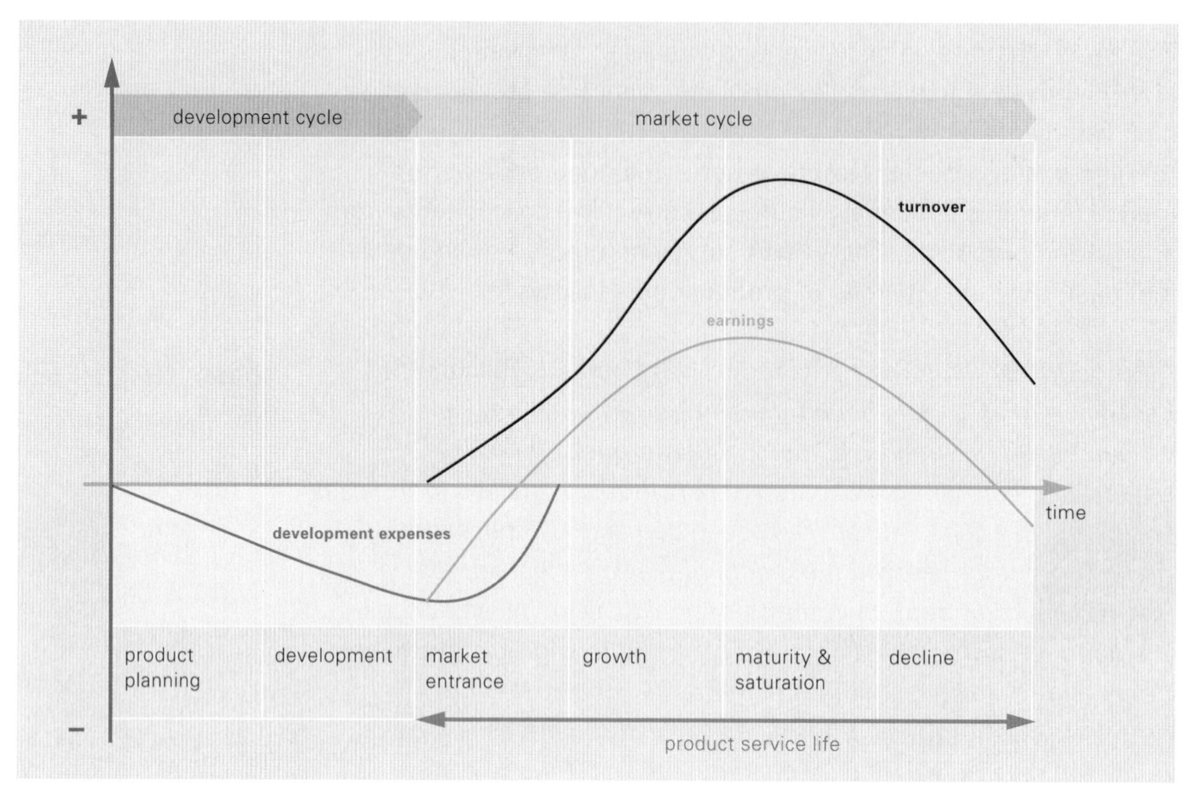

Figure 2.2 Model of a product life cycle

The higher the development expenses for a product become, the more damaging a late market entrance becomes. For this reason the automotive industry started early on making their manufacturing plants more flexible with the result of customized vehicles leaving the conveyor belts today. Although this idea would have been considered as futuristic in the 50s, it was realized by strategic use of efficient automation technology components.

2.2 Developmental Stages of Grippers

At an early stage the idea of offering complete unit construction systems and feeding technology, robots, and grippers, for automation technology was of major importance in order to be able to flexibly react to Mechanical Engineering demands. Consequently, the first gripper modules were developed as standard products as early as 20 years ago.

By comparing three parallel jaw grippers at various stages of development, the efficiency increase can be illustrated (see figure 2.3). The standard gripper PPG offered by SCHUNK in 1983 already had very good ratios at the time. The force/weight ratio, i. e. the gripping force in relation to the weight of the gripper multiplied by the stroke of the finger, was at 2.4 J/kg for the short stroke. The next milestone was set by the PGN gripper which was built on the same functional principle as the PPG. This PGN was able to reach a energy/weight ratio of 7.3 J/kg. The following generation of this successful gripper series significantly increased the energy/weight ratio to 8.1 J/kg while its service life was improved at the same time.

$$C = \frac{F_g\, s}{G} \quad \left[\frac{Nm}{kg}\right]$$

F_g = gripping force [N], s = stroke [m], G = weight [kg]

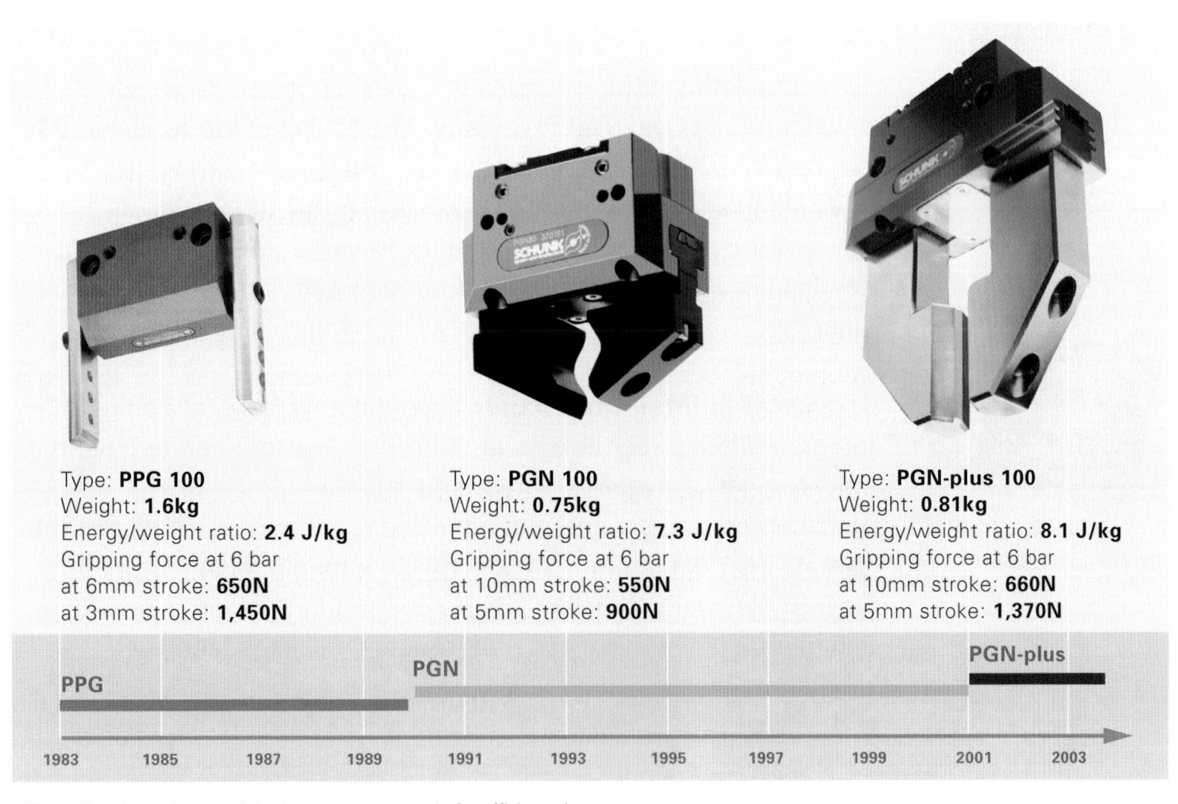

Figure 2.3 Comparing parallel grippers as an example for efficiency increase

Old and new: Angular gripper (left) and parallel grippers (right)

Gripping hands are a good example for explaining the stages of gripper development as mechanical innovations as well as innovative sensor and drive technology are included. The hand is the most flexible tool human beings have at their disposal. At the beginning of human evolution, however, the hand was not equipped with the fine motor skills which allow us to work with keyboards, writing equipment or other current tools. It took millions of years to train such refined movements.

Gripping tools are undergoing a similar evolutionary process. Grippers with a secure grip (force- or form-lock) were the first needs of the moment in automation technology. Increased manufacturing flexibility required improved gripper flexibility.

Two completely different types of "artificial hands" have been developed over the past few years. The so-called **modular hands** can be combined with kinematics and movement facilities. They include all components necessary for function, such as actuators or sensors. Modular hands are larger in size than human hands because the actuators have to be integrated into the hand kinematics.

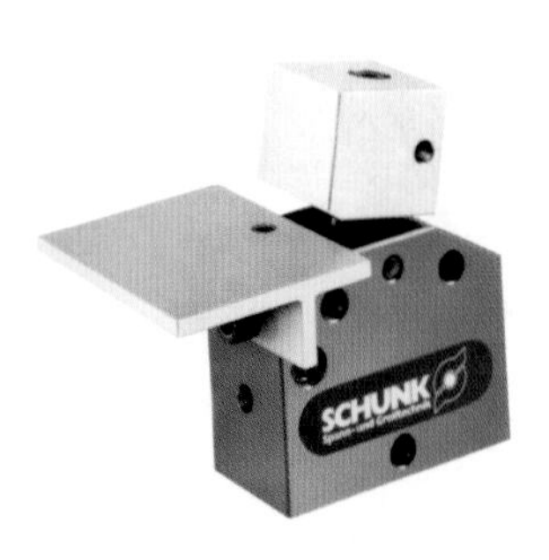

1-finger gripper

The so-called **integrated hands** have the advantage of being integrated into a robot arm and, therefore, do not need to have the actuators (drive systems) integrated into their housing. The actuators are mostly outsourced to the robot arm which allows the use of larger actuators with a relatively strong gripping force. Nevertheless, transferring drive forces to the gripping fingers over a longer distance is difficult and frequently causes technical problems.

types of artificial hands

modular hands

- are adaptable to any kinematics
- include all components required for function (sensors, actuators, ...)
- are larger than human hands due to the size of current actuators
- have a lower gripping force than integrated hands
- require overall complex design

integrated hands

- hand is integrated into the robot arm
- components, e. g. actuators, can be outsorced to the arm
- larger actuators produce higher gripping force
- transfer of forces to the fingers or joints is difficult

The first modular hands were the Stanford Hand and the Barret Hand.

Stanford/JPL Hand

Hand/arm integration: **Modular**
Abilities: **Internal manipulation**
Number of fingers: **3**
Number of links: **10**
Number of joints: **9**
Degrees of freedom: **9**
Palm: **No**
Size compared to human hand: **Equal**

Sensors:
- rotary transducers in each motor
- strain gauge sensors
- tactile sensor array at the finger tip joints (8x8)

Reference:
Salisbury, Stanford University, 1983

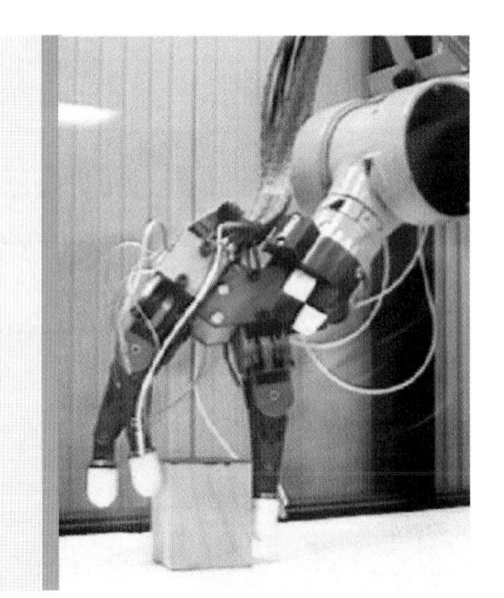

The **Stanford Hand** was built in 1983 and is equipped with tactile sense contacts on the fingers which are to imitate the human sense of touch. The gripper was equipped with just three fingers but could still manipulate the workpiece in its hand.

Barret Hand

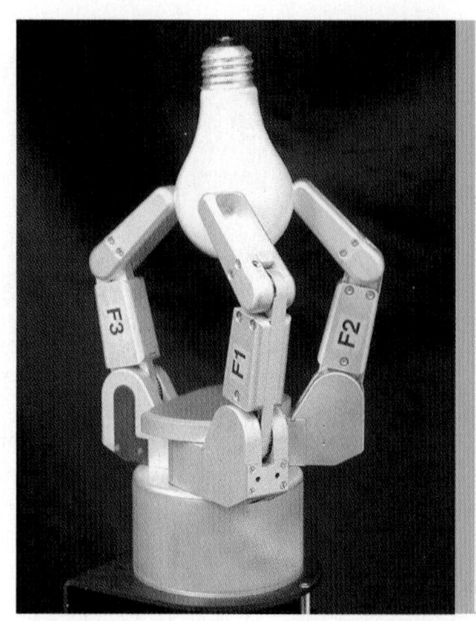

Hand/arm integration: **Modular**
Abilities: **Gripping**
Number of fingers: **3**
Number of links: **9**
Number of joints: **8**
Degrees of freedom: **4**
Palm: **Yes**
Size compared to human hand: **Equal**

Sensors:
- optical rotary transducers in the motors
- strain gauge bridges at the finger tips

Reference:
W. T. Townsend, Barret Technology, Inc., 1988

The **Barret Hand** has been constantly improved over the past years and is currently marketed and distributed by SCHUNK in Germany. In 1988 it was already introduced as a modular gripper and its functionality and applications are detailed in Chapter 3.

Robonaut Hand

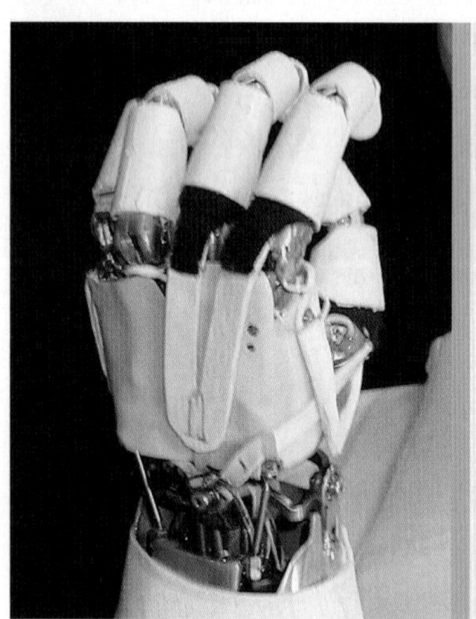

Hand/arm integration: **Integrated**
Abilities: **Internal manipulation**
Number of fingers: **5**
Number of links: **22**
Number of joints: **22**
Degrees of freedom: **14**
Palm: **Yes**
Size compared to human hand: **Equal**

Sensors:
- precise position transducers in the joints
- rotary transducers in the motors
- tactile sensors (being developed)

Reference:
C.S. Lovhik, M.A. Diftler; NASA; 1999

The **Robonaut Hand** is one of the first to be designed for applications in space. According to NASA the highest demands on the materials used, such as extreme temperature resistance, distinguish this hand from all hands produced so far. Even eventual gas emissions by the hand and their influence on other space systems were taken into consideration.

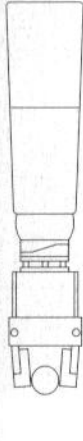

The Robonaut Hand was built on the basis of integrated kinematics and is very human-like in size and mobility with its more than 14 degrees of freedom. It is able to use a screwdriver and even grasps small objects with a pair of tweezers.

Hand/arm integration: **Modular**
Abilites: **Internal manipulation**
Number of fingers: **4**
Numbers of links: **18**
Number of joints: **17**
Palm: Yes
Degrees of freedom: **13**
Size compared to human hand: **Larger**

Sensors:
- potentiometer in the joints
- rotary transducers in the motors
- 6-axis force-torque-sensor (finger tip)
- torque/moment sensors in each joint

Rererence:
Butterfass, Hirzinger et al; DLR; 2004

DLR Hand

The DLR Hand II by the DLR Institute of Robotics and Mechatronics

Compared to the NASA Robonaut the **DLR Hand** by the German Aerospace Center operates with one finger less. Included in the 4-finger hand and adaptable palm, are 13 degrees of freedom and nearly 100 sensors.Cable pulls and the strong-reduction and low-friction DLR spindle, for which worldwide patents have been granted, enable the entire actuation to be integrated into the hand.

The goal of imitating the human hand in its flexibility and efficiency has not been achieved by one of the above "artificial hands". Grippers will only be available as "specialized hands" for some time. This also applies for modular grippers in hand prosthetics. A precision gripper for manual operations is offered by Otto Bock, Duderstadt in Germany. It uses electronic drive technology and weighs about 500g. Thus, it can be replaced but appropriate special hands have to be used for the respective tasks.

(Source: Otto Bock, Duderstadt, Germany)

Other gripping technology modules have gained considerably in functionality, quality, and sensor integration. The increasing number of application fields are a clear sign of this development.
The broader the range of applications and the greater the required flexibility of the respective gripper, the more it usually costs if such a gripper module has to cope with several products.

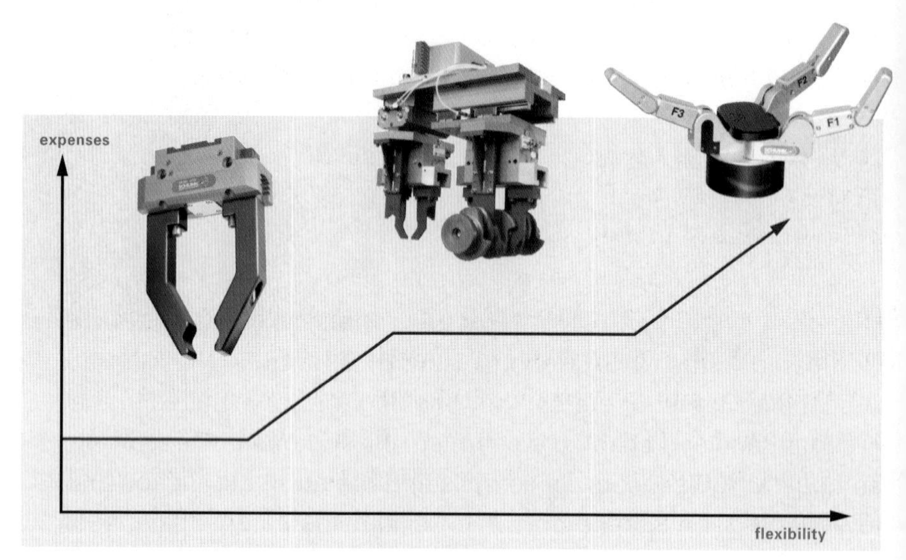

Figure 2.4 Expenses/flexibility ratio for gripper applications

Figure 2.4 clearly shows this fact. Naturally, for many automation tasks a simple but relatively unflexible gripper will be sufficient. For more demanding applications, a special construction, possibly in combination with standard grippers, is required. Only applications which do not allow the gripper to be changed and have to deal with numerous different workpieces make a highly flexible gripper solution a necessary investment. Special solutions currently on the market are close to their efficiency limits in relation to payload and velocity. As a result, "artificial hands" are mainly used for service robots and in Research & Development today.

2.3 Robot History

Gripper technology and "robot revolution" nearly go hand in hand. The first years were characterized by a euphoria which were curbed by practical drawbacks.

The term robot is derived from the Czech word "robota" which was used to describe the part of the serf, the hard-working slave, or submissive servant in the theater premiere of Rossum's Universal Robots (R.U.R.) by Karel Capek (1890-1938) in 1921. During that time period, various terms for mechanical machines were in use. Words such as simulators, automats, rational machines and others were circulating. Since then humankind has come closer to the idea of eliminating sometimes dangerous physical work. Only the conception of the robot as a universal "slave" taking over all of the work remains outdated.

The beginning of the 50s was the start of stationary robot development. The brainchild of George Devol and Joe Engelberger, Unimate, was a robot weighing two tons and was controlled by a program stored on a magnetic drum. Unimate was first installed at General Motors in 1961. Henceforth the automotive industry has been and will continue to be the driving force of industrial robot development.

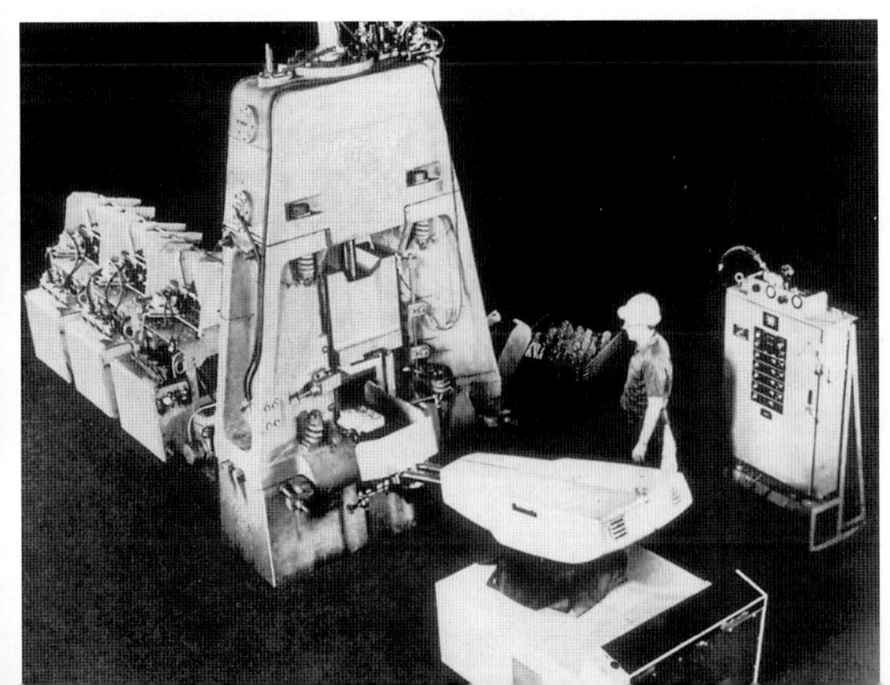

Unimate feeding a press (source: Fraunhofer IPA)

So-called "human" robots were already one of the goals of early robotics. In 1963, researchers at Rancho Los Amigos Hospital in Downey, California, constructed the Rancho Arm for the support of physically challenged people. At the Massachusetts Institute of Technology (MIT) in 1968, Marvin Minsky developed the Tentacle Arm with twelve joints designed to reach around obstacles Victor Scheinmann, a Mechanical Engineering student working in the Stanford Artificial Intelligence Laboratory (SAIL), developed the Stanford Arm in 1969. This 6degree of freedom (6-dof) all-electric mechanical manipulator was hardly a human-like hand but one of the first "robots" designed exclusively for computer control and micro surgery. Projects included the assembly of a Model A water-pump in 1974 and this is how the "arm" development found its way into the automotive industry.

Syntelmann II
Electric tele manipulator with 9 degrees of freedom per arm, position- and force-controlled, sensors for forces, sounds, temperatures (in front); operator with exo-skeleton transducer system, force feedback system, and stereo image transmission system (in the back).
(Source: K. H. Dröge)

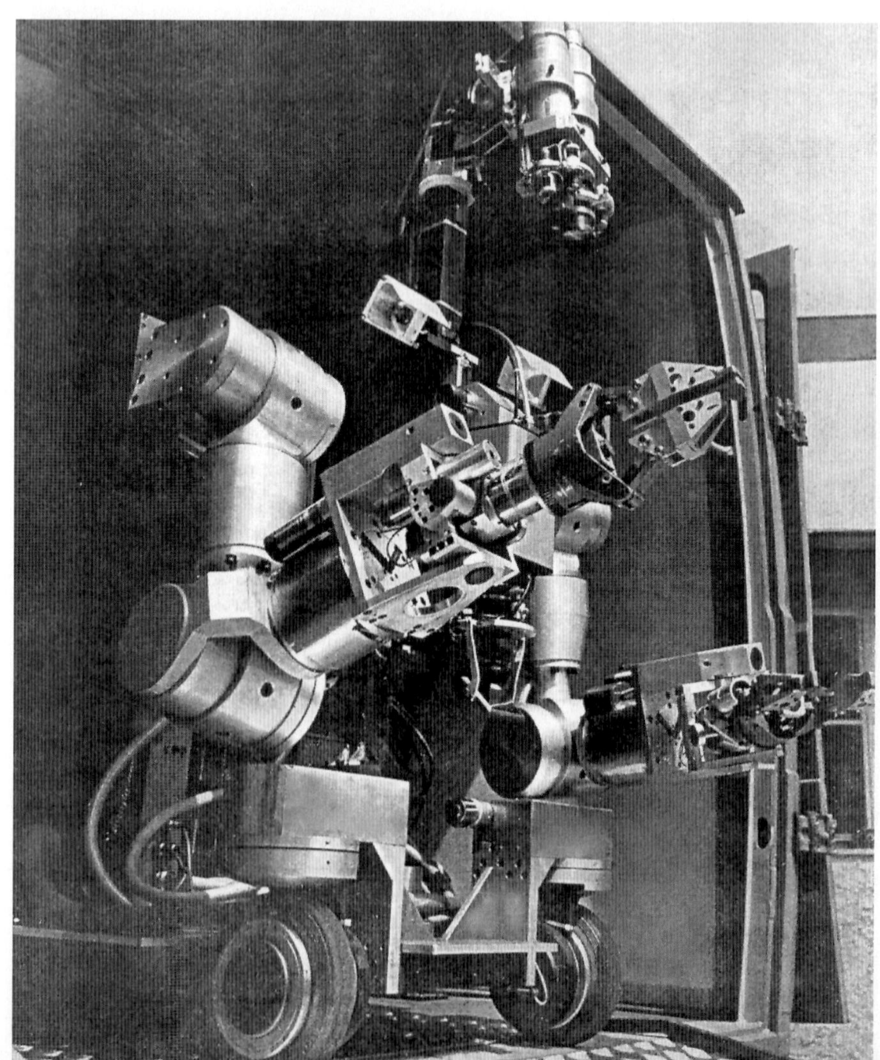

The "Syntelmann" was developed by Kleinwaechter in Freiburg, Germany, parallel to the development of the Stanford Arm in the U.S., but did not withstand the test of time.

Syntelmann was commissioned by the German Ministry of Education and Science to make repairs in case of nuclear power plant catastrophies or to save humans in contaminated areas. The name Syntelmann was used as an abbreviation for "Synchron-Tele-Manipulator". The manipulator was equipped with two hands, one for handling a payload of 25kg and the other for performing high-precision operations. A precondition for millimeter precise positioning of workpieces were precision drives and angle transducers which powered the arms.

On the basis of the Standford Arm, the Programmable Universal Manipulator for Assembly (PUMA) was developed and a version of this robot arm licensed by Unimation began working at the General Motors Technical Center in 1978.

The first commercially available micro-computer-controlled robot named T3 (The Tomorrow Tool) was developed by Richard Hohn for the Cincinnati Milacron Corporation in 1973. Built as a robot with a so-called Computerized Numerical Control (CNC), the first type of the T3 was hydraulically powered and not available on the market until five years later in 1978.

ASEA introduced its first electronic-control robot in 1974. Weighing 125kg, the IRB 6 was able to move payloads up to 6kg. It approximately cost $80,000 and managed to cope with 16 electric inputs.

Once the initial obstacles of robot control were crossed numerous variants and component series followed, mainly differing in payload capacity and workspace. The IRB 60, for example, managed a payload of 60kg.

IRB 6 by ASEA with 6kg payload capacity (source: ABB)

IRB 60 by ASEA with 60kg payload capacity (source: ABB)

Meanwhile every larger robot producer offers a broad range of robot kinematics for various needs. These kinematics and its variants are detailed in Chapter 4. At this point we are concentrating on comparing renowned robot producers and their products today and 30 years ago. A direct comparison of kinematics and its controllers shows a clear trend: Major progress has been made in drive and control technology as well as in software for robots, i. e. developments which are not always obvious at first sight.

Special kinematics were developed for handling presses in order to significantly increase the cycle time of robots. The Bilsing-Unimate, which you can see in the picture, is a good example of a highly individual solution which can hardly be used for any other purpose. Limited application was responsible for uncompetitive prices with the result that standard kinematics are mainly used for press handling today.

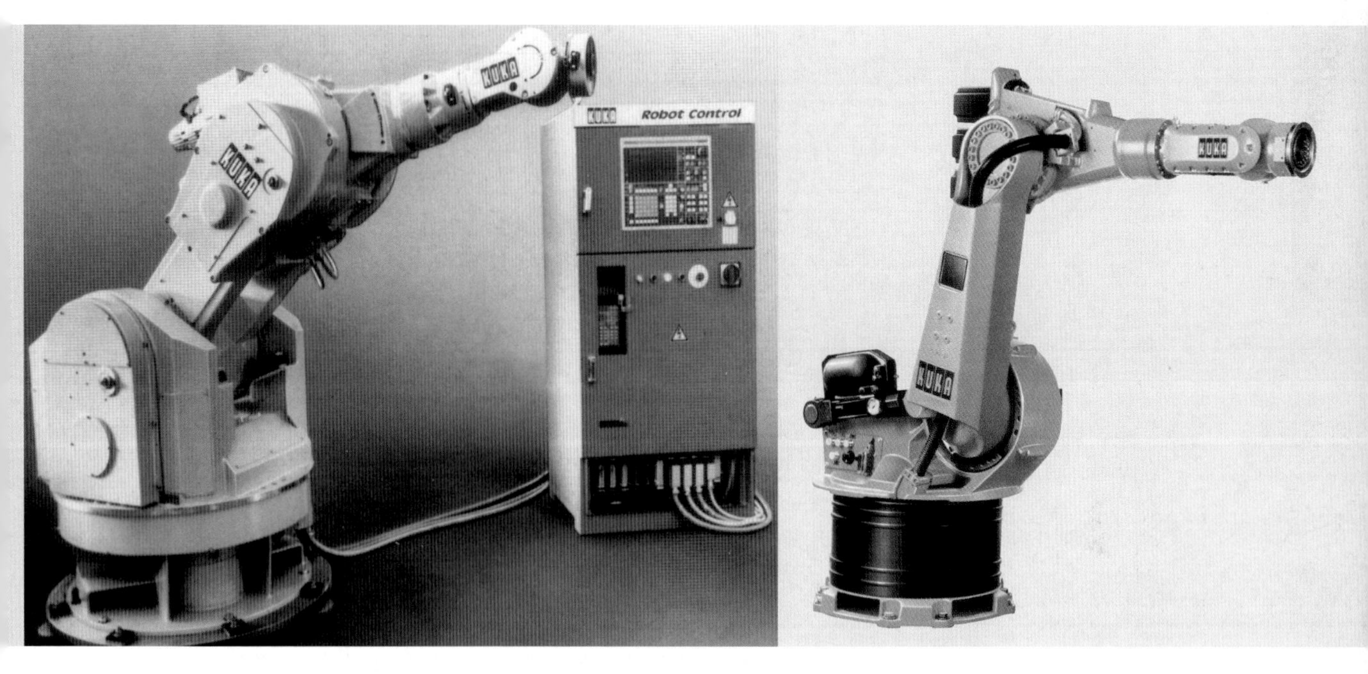

Various robot generations by KUKA (source: KUKA)

Bilsing Unimate robot for press handling (source: Fraunhofer IPA)

In 1984 the MBB VFW managed a regular payload of 50kg up to a maximum 200kg while it weighed a solid 2,350kg. The approximate list price of $165,000 compared to an industrial worker's $10,000 labor costs (incl. ancillary wage costs) per year. Looking at these power and price levels it is obvious that robot producers were hardly able to sell their products.

At the same time the ROBOT 625 by Reis Obernburg had the same kinematic principle as the MBB-VFW. The ROBOT 625 only weighed 750kg at a regular payload of 25kg, a clear improvement on the weight/payload ratio. Even with its 64 inputs and 32 outputs it exceeded the MBB-VFM by the factor 4. In addition, it offered a significantly larger workspace and at $80,000 cost less than half the price.

MBB-VFW robot with controller (source: Fraunhofer IPA)

As manufacturers then could not offer more than one to three different housing and payload categories it was essential for them to find the right applications for their kinematics. This situation has not changed much, with the difference that manufacturers now offer a broad range of housing and payload options and, therefore, are able to meet almost any application.

ROBOT 625 handling motor blocks (source: Fraunhofer IPA)

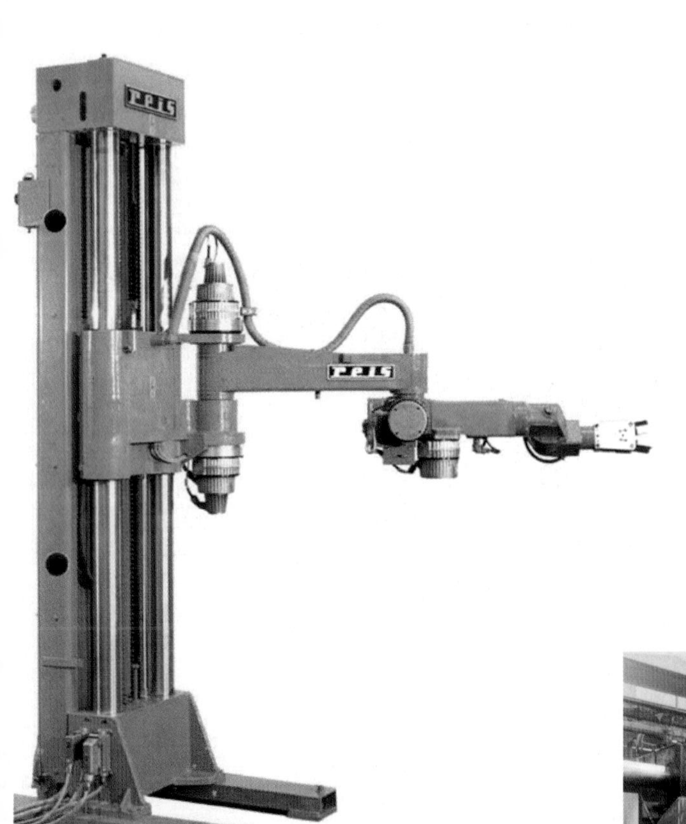

ROBOT 625 (source: Reis)

ROBOT 625 (source: Reis)

All robot producers try to add new applications to their key applications as shown by the figures on plans for the VW robot application. Bosch first used the SCARA SR 800 for internal purposes while current systems are used for the most diverse assembly tasks.

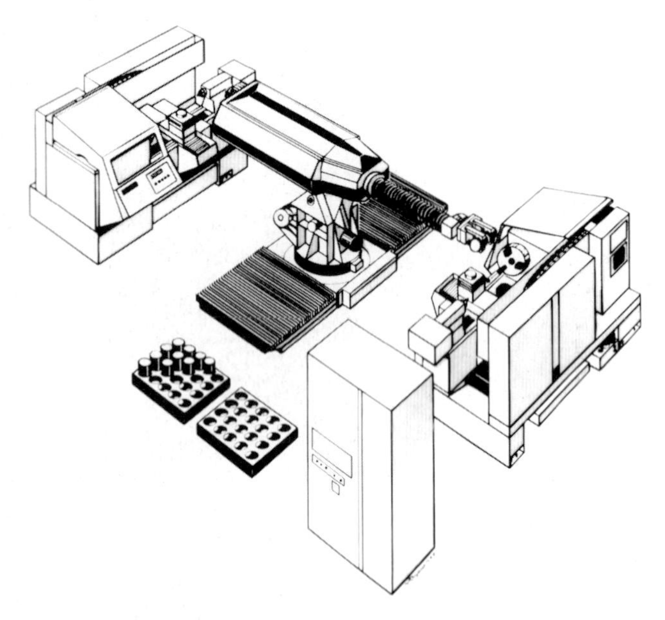

Plans for the use of a VW robot for feeding tooling machines (source: Fraunhofer IPA)

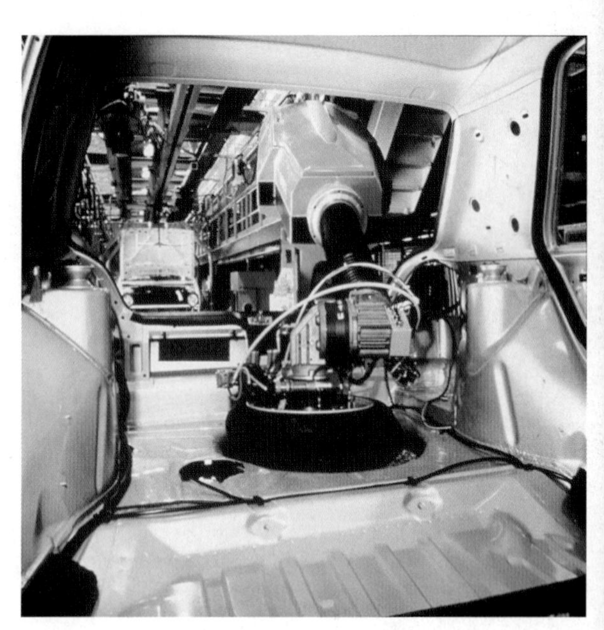

Typical application of a VW robot for inserting the spare wheel into the Golf II (source: Fraunhofer IPA)

Left:
BOSCH SCARA series SR800 at a double belt transfer system (source: Bosch Rexroth)

Right:
Current SCARA-Roboter SR 8 (source: Bosch Rexroth) now sold by Stäubli

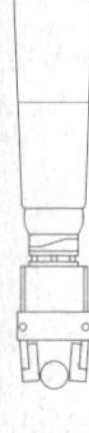

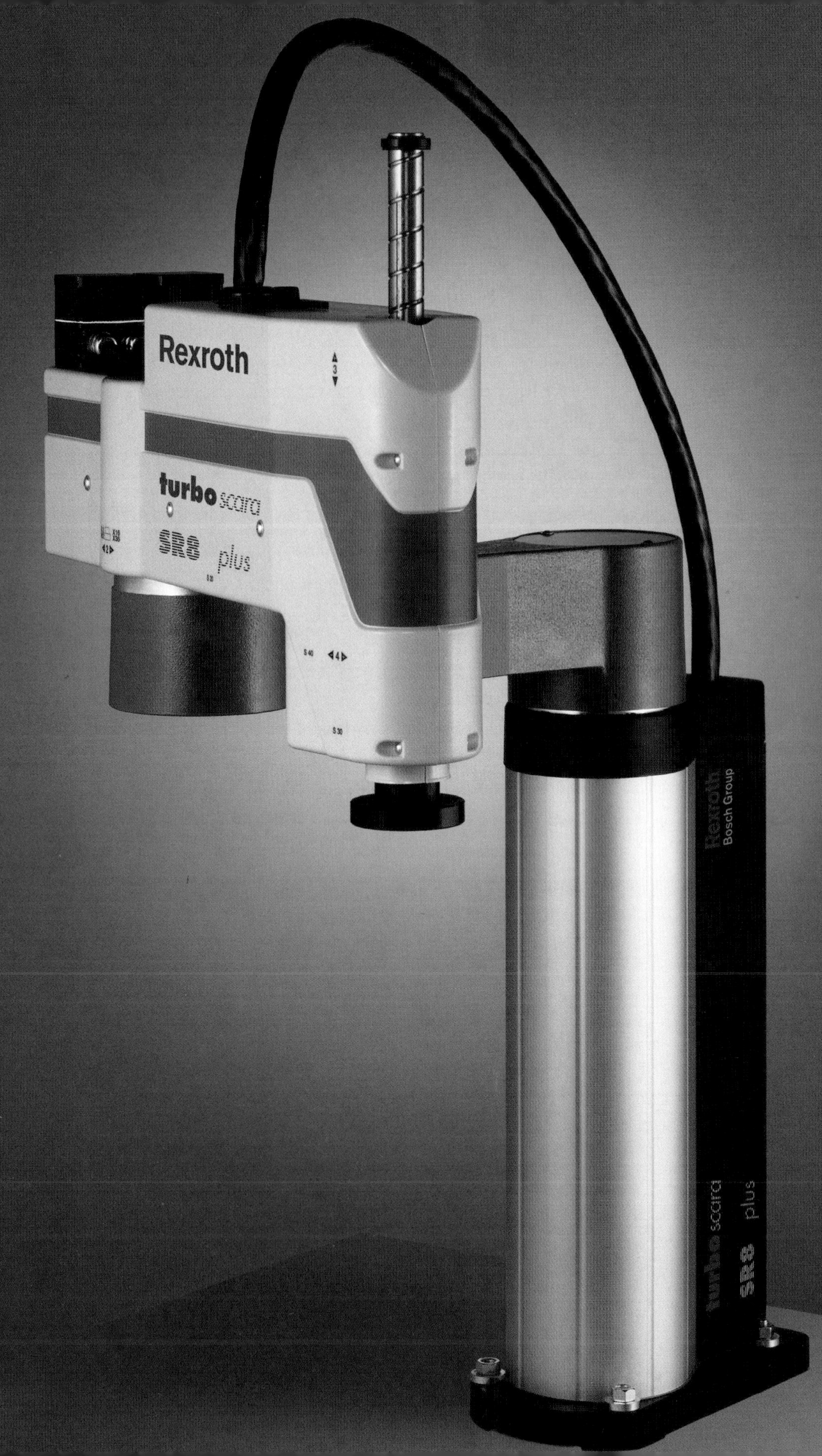
Rexroth
turbo scara
SR8 plus
Rexroth
Bosch Group
turbo scara
SR8 plus

Just a few robot producers survived the stiffening competition during the first years. In the first robot catalogs published in the former German Democratic Republic (1983 edition by the Forschungszentrum des Werkzeugmaschinenbaus, Karl-Marx-Stadt) and nearly parallel in the Federal Republic of Germany (1984 edition by the Fraunhofer Institute for Manufacturing Engineering and Automation IPA in Stuttgart, Germany) all robot procucers and their products are listed.

The 1984 Fraunhofer IPA catalog names approximately 80 producers while a much lower number appears in the AUTOMATICA 2004 Munich Germany exhibitors directory. Although the AUTOMATICA 2004 fair just started in 2004, the reduced number of German robot producers is clearly visible. After 20 years, only five out of 35 German robot producers listed in the 1984 Fraunhofer IPA catalog are present at the AUTOMATICA 2004.

Producers such as Pfaff Industriemaschinen or Jungheinrich were two of the pioneers, just like large enterprises such as Siemens or Volkswagen. However, most of the smaller robot producers simply could not cope with the target quantities.

A complete overview is bound to exceed the volume of this book. The photographs and figures illustrate the impressive number of different companies in Germany which were engaged in the production of robots.

Robots initially started out in the U.S. but today's world production is mainly situated in Japan, Sweden and Germany. Fast growing markets in China and India are setting out to enter the market with their own products. Major Japanese companies building robots are Yaskawa (also known under the name MOTOMAN in Germany), Kawasaki and Fanuc. Renowned brands for small robots are EPSON, Mitsubishi and Hirata.

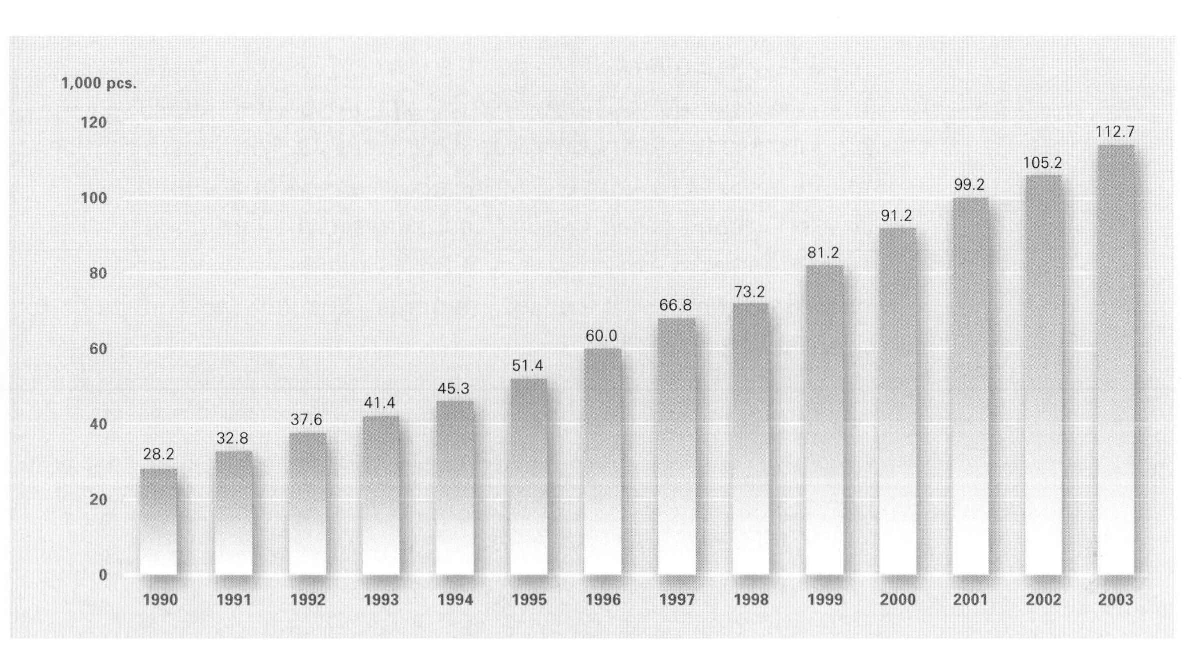

Figure 2.5 Industrial robots in Germany – installations in use, adjusted according to the IFR method (source: VDMA)

The number of robots in use and their growth rates are regularly recorded by the VDMA. The 2003 World Robotics survey, published by the IFR (International Federation of Robotics) in cooperation with the United Nations Economic Commission for Europe (UNECE), counts a total 112,693 active (i. e. operating) robot installations. The IFR/UNECE adjusted statistics exclude robots with a service life of more than 12 years or their related applications according to their expected average service life.

In the past few years the robot sellers` market clearly concentrates on a few large producers who are direct suppliers to the automotive industry. Some of the manufacturers were able to strongly increase sales due to their concentration and higher sales at the end of the 90s.

As shown in figure 2.6 the first 15 years of robot production at KUKA accounted for 12,000 robots as compared to 48,000 produced in the years between 1996 to 2003. Four times as many robots were built and sold within about a third of the time. This enormous growth rate is connected to the introduction of the first PC based robot control in 1996. PC technology created new opportunities for sensor integration and ideal preconditions for user-friendly applications.

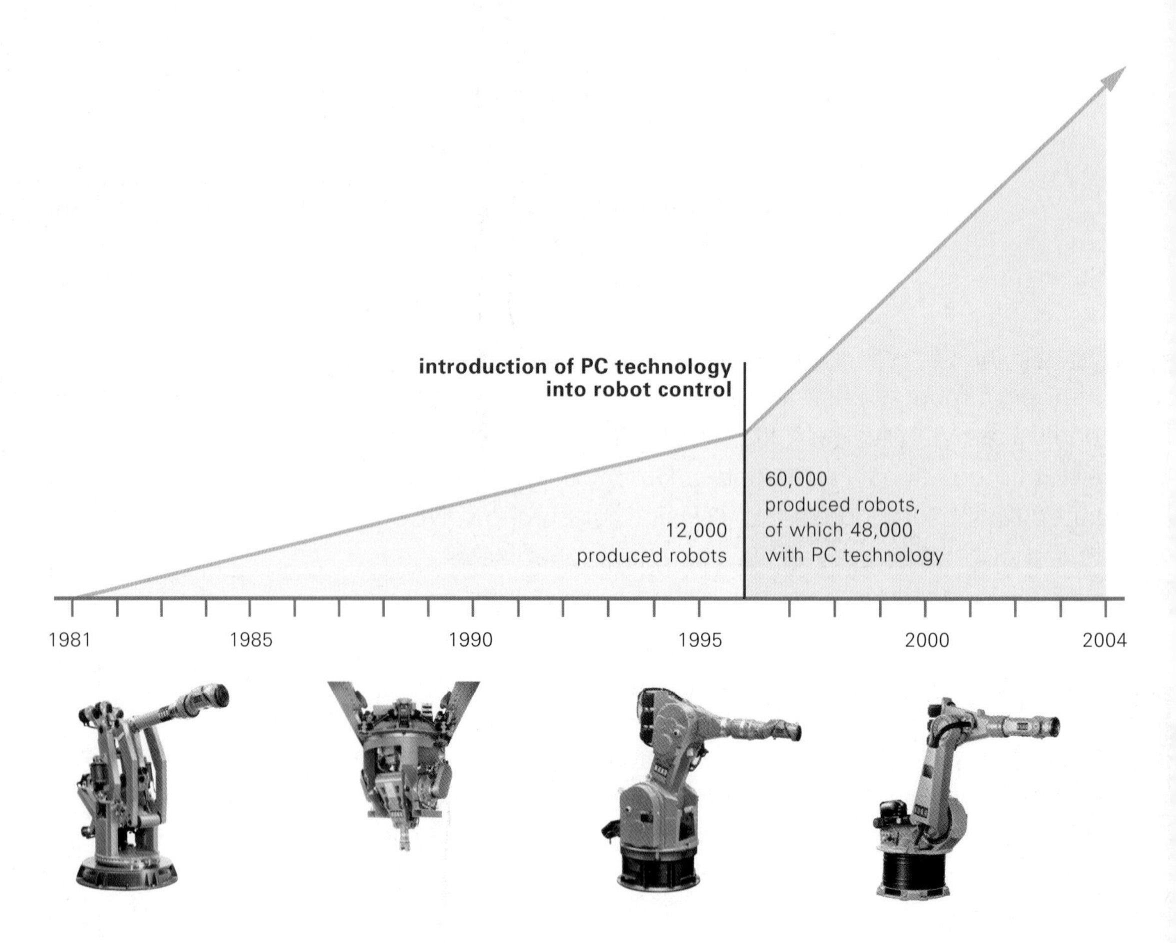

Figure 2.6 Development of industrial robot technology (source: KUKA)

Right:
Robot control unit in 1982
(source: Fraunhofer IPA)

FREIGABE
A.S.R.SERVOTRON

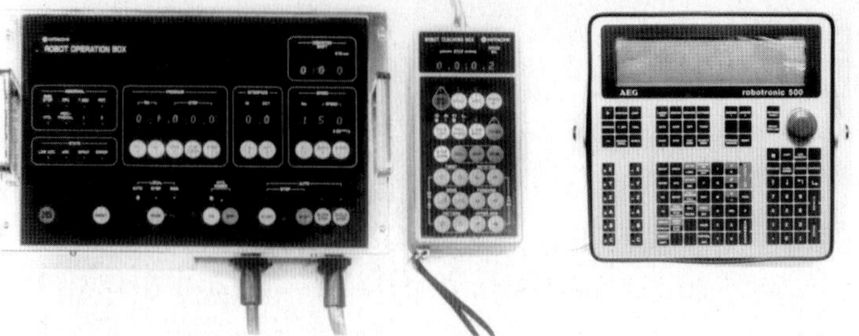

Former machine/operator interface (source: Fraunhofer IPA)

In terms of user-friendly machine operator interfaces, enormous improvements have been made which are illustrated by some examples of robot programming devices.

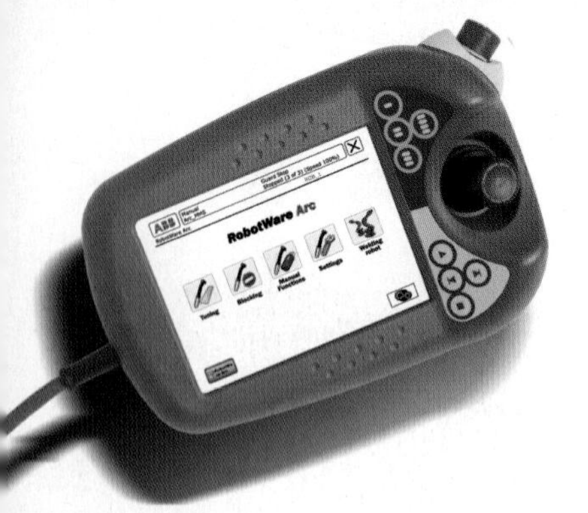

Modern machine/operator interface (source: ABB)

Modern machine operator interfaces particularly show a trend towards user-specific interfaces which can be customized to meet individual requirements. Significantly it can be seen the reduktion of hardware switcher and better graphical possibitlities.

The dynamic development of robotics is depicted in figure 2.12. Significantly it is visible that the productlife of a robot has declined also over the last years.

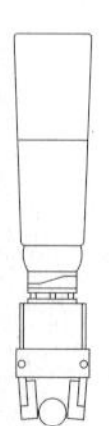

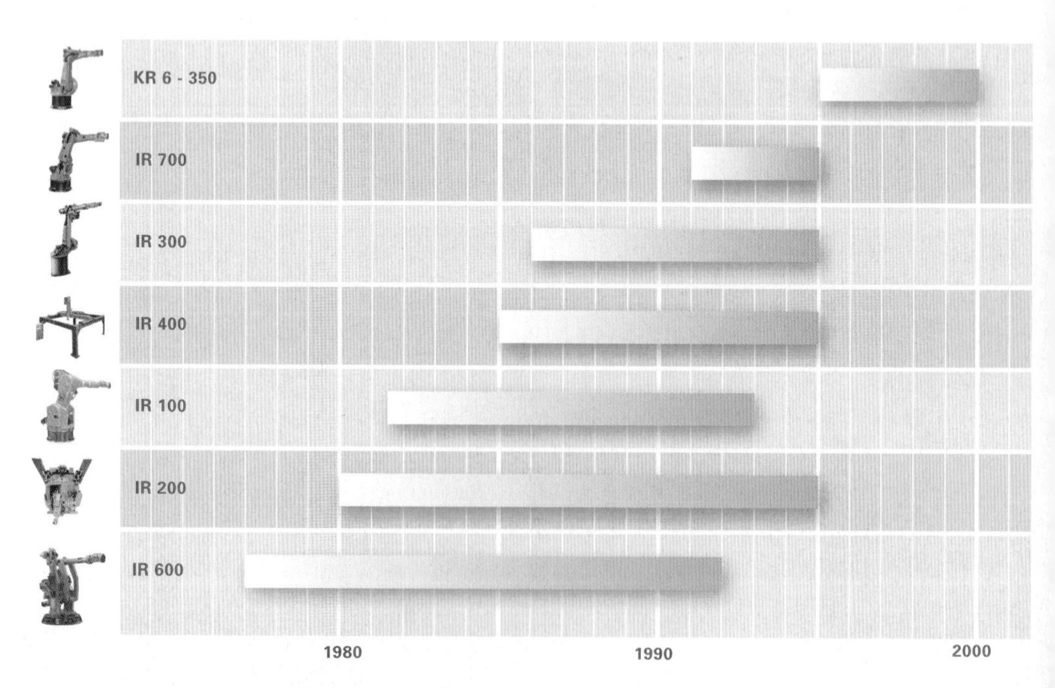

Figure 2.12 Different series of one robot producer (source: KUKA)

In order to enable the flexible use of robots, it is necessary to let them know more about their environment. This is achieved by either programming environment conditions beforehand or else gaining information through sensors and transmitting this information to the robot. Coordinates of movement, velocity and acceleration for workpiece handling can be influenced by information from the environment. At an early stage research concentrated on sensors similar to the human eye to grasp situations and components for further processing, leading to so-called image processing systems.

Camera system in the 70s (source: Bosch)

The development of camera and computer technology for image processing is of great importance for flexible robot applications. The tables show a so-called tele sensor introduced by Bosch at the beginning of the 80s as well as a modern "intelligent" camera system with integrated picture evaluation. The tele sensor was sold for a mere $36,000 which was a bargain at the time. Only one binary picture could be processed at a time when evaluating the recorded pictures. Thus,applications were subject to light interferences and did not allow for smooth processing. As a result the conveyor belt had to be stopped to take the picture.

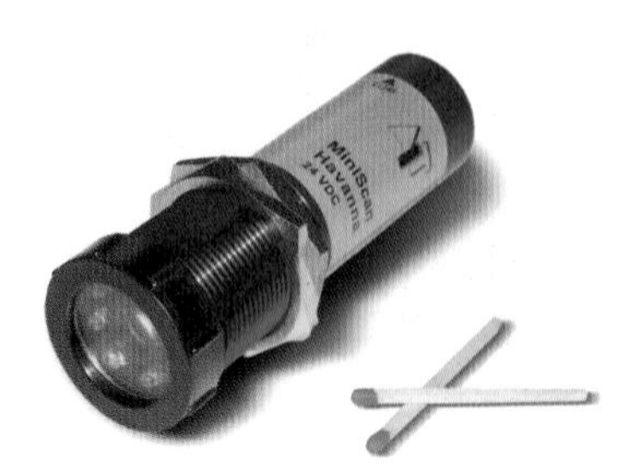

Camera system in 2004 (source: AIT Göhner)

Two monitors were integrated, one for the programming device and the other for showing the picture taken by the camera.

Modern cameras and evaluation systems are able to evaluate 110 pictures per second at minimal costs compared to the first systems (ca. $4,200 incl. lighting and lens for the system as shown). The first efficient image processing cards for PCs were available in the order of $50,000 during the early 90s. Interface problems, which were an initial hurdle for the simple integration of sensors into a robot control, are nearly solved by a built-in evaluation computer as shown for the camera system in the picture. An evaluation of various shades of gray or even colors enables inspection of complex structures and geometries.

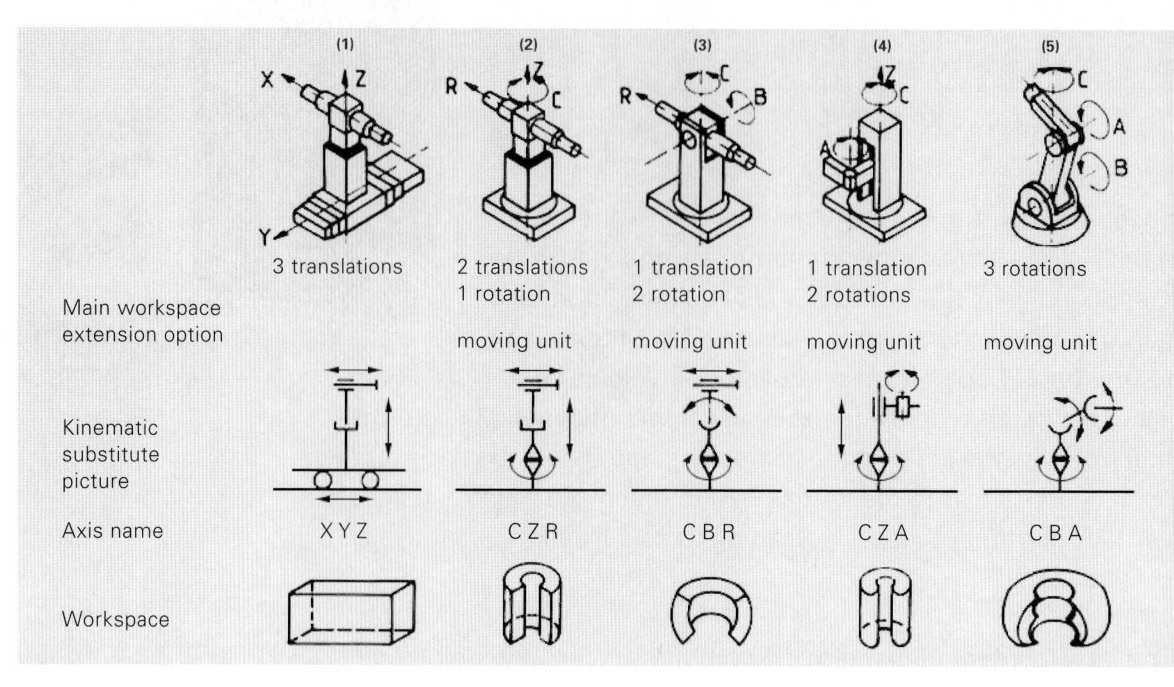

Figure 2.7 1983 regular kinematic types and their workspaces

Camera technology is an essential part of the quality control of workpieces. Image processing has developed into a robust and easily accessible technology as sufficient piece numbers of these sensors are currently offered on the market. In addition, camera technology is directly connected to the digital camera mass market, which results in favorable prices at parallel development boosts on a yearly basis. While a 256 x 256 pixel standard used to be available, today's industrial standard is 1300 x 1024 pixel.

Within the next years industrial evaluation cameras will reach a 2000 x 2000 pixel standard, which again allows evaluating and measuring workpieces with even higher precision.

2.4 Robot Statistics

Higher flexibility and efficiency enabled robot technology to conquer more and more applications in numerous fields. The quantity record speaks for itself as it documents robot technology distribution.

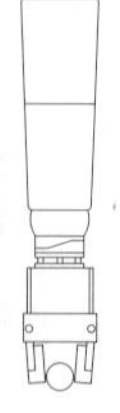

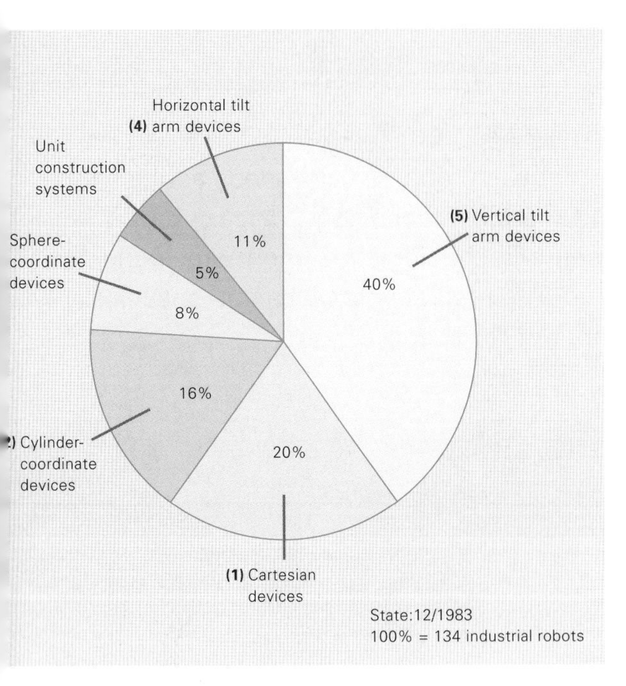

Figure 2.8 Proportion of kinematic types distribution (source: Fraunhofer IPA)

1983 statistics show the distribution of robot types used in Germany (figure 2.8). Basis of the statistics were 134 three-axis robots. In comparison, the figures 20 years later are quite much more impressive: In 2003 an overall 2,522 three-axis robots were statistically registered by the VDMA; an overall 9,040 robots with six axes had already sold in Germany alone.

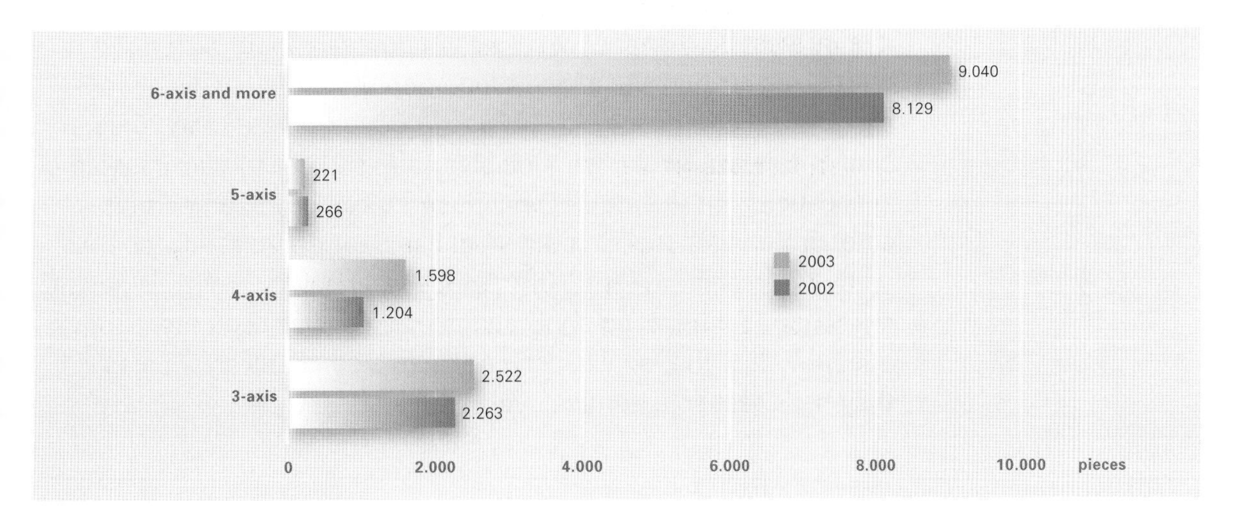

Figure 2.9 Development of the German robotics market in relation to the number of axes (source: VDMA)

If you consider the fact that each robot usually needs additional peripheral devices such as feeding technology, magazines, grippers, sensors, and safety technology, the enormous market volume becomes obvious. The market volume can be specifically determined by the number of robot applications.

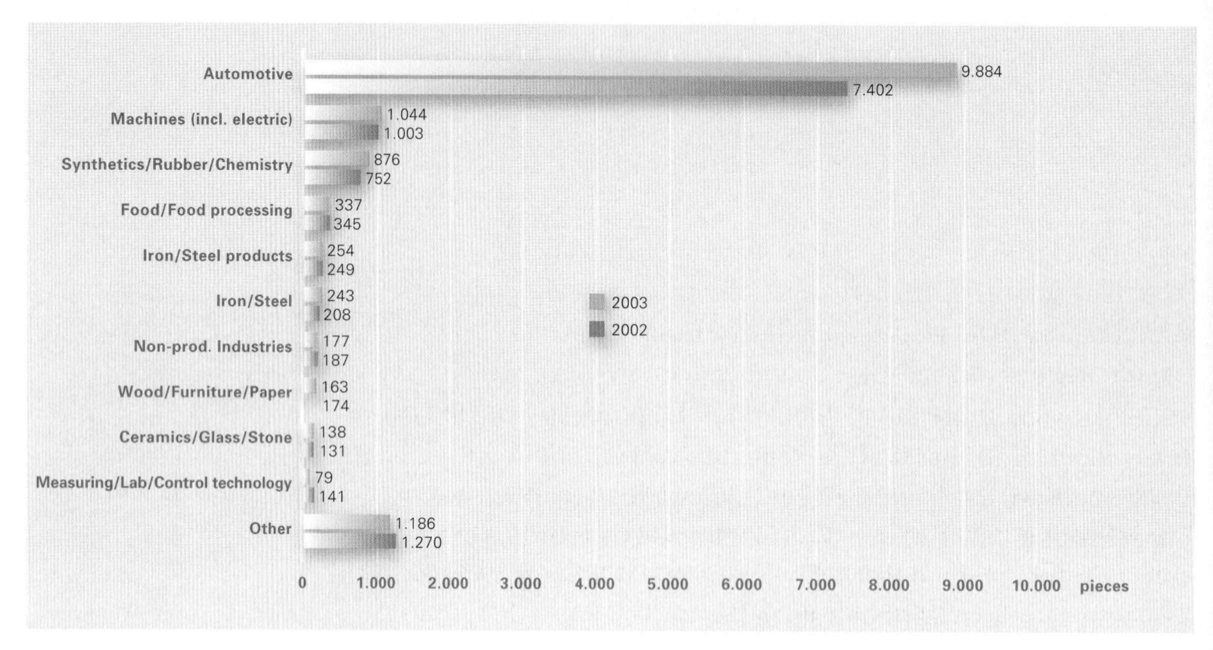

Figure 2.10 Industrial robot applications according to industries 2002 and 2003 (source: VDMA)

It is interesting that applications for industrial robots with three or more axes account for a high proportion of robots handling workpieces. The great rise from 2002 to 2003 with more than 30 percent in this category is remarkable. As this increase was initiated by applications in the automotive industry it can be assumed that an enormous rationalizing potential has been successfully opened up by the automotive industry. Gripper technology, which is normally used for handling applications, has especially profited from this growth. Other applications came to a standstill as soon as they had reached a certain level.

The German Federation of the Engineering Industries VDMA gives the following outlook in its yearly published statistics for the year 2004:

"In 2004 a further increase in robot piece numbers is expected for the German market as well. The automotive industry will remain the largest customer with substantial investments in 2004. Rubber and synthetics producers, the chemical industry, machine (incl. electric) producers, and the metal manufacturing industry will have to increase their investments accordingly. The potential for using robots in the food processing and the packaging industry is obvious as the demand for automated solutions is high. The furniture industry is another potential customer as it it forced to reduce costs of production, too. Higher payloads, higher dynamics, sensor technology, network technology (of communicating robots), and image processing systems, offer increased application options for so-called "intelligent robots" or "multi-robots". Robot technology faces another quantum leap. In high-wage countries, the unit labor costs can be effectively reduced by automated solutions. Technical concerns of some industries, such as food processing, are answered by user-friendly systems. The food processing industry in Germany still offers great potential for the use of robots and automated systems."

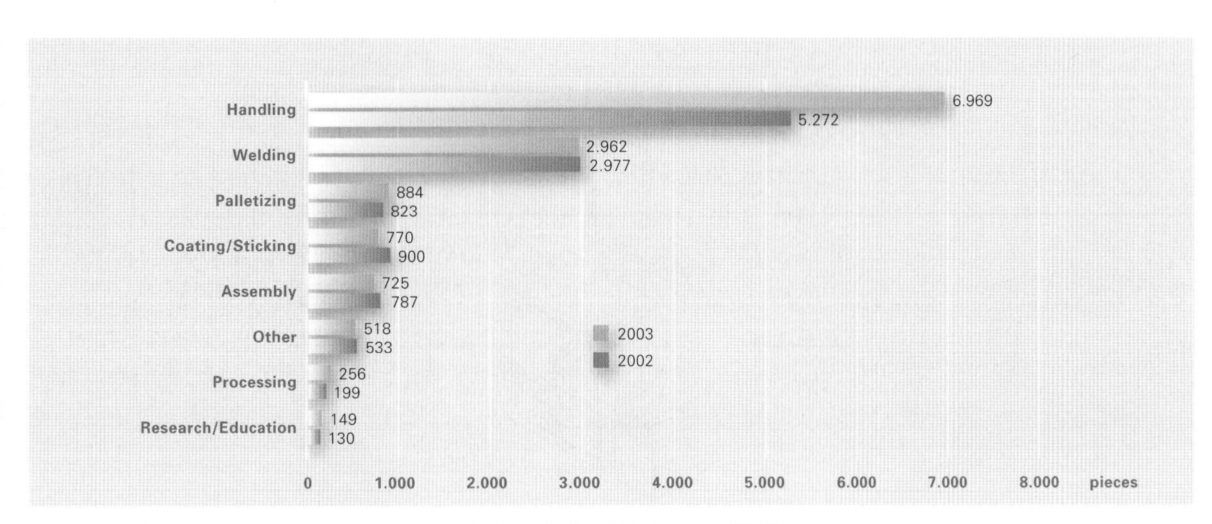

Figure 2.11 One- and two-axis moving modules – applications 2002 and 2003 (source: VDMA)

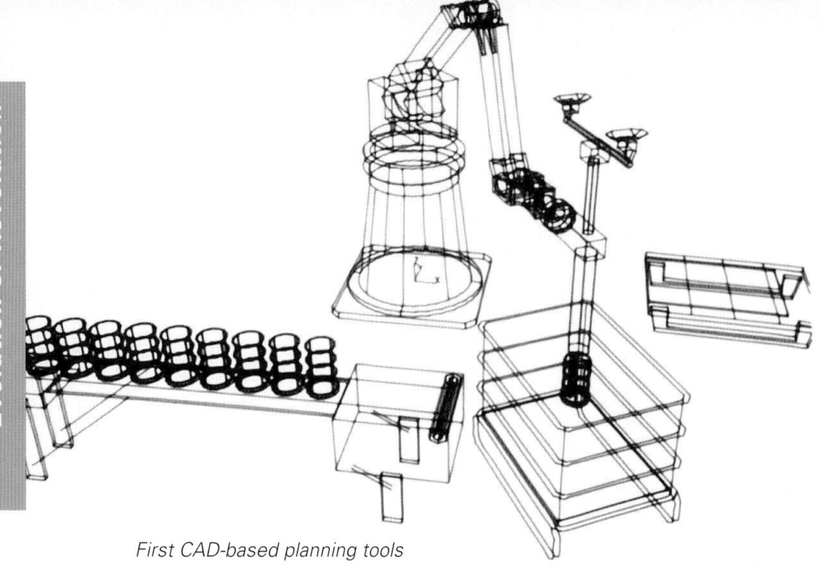

First CAD-based planning tools

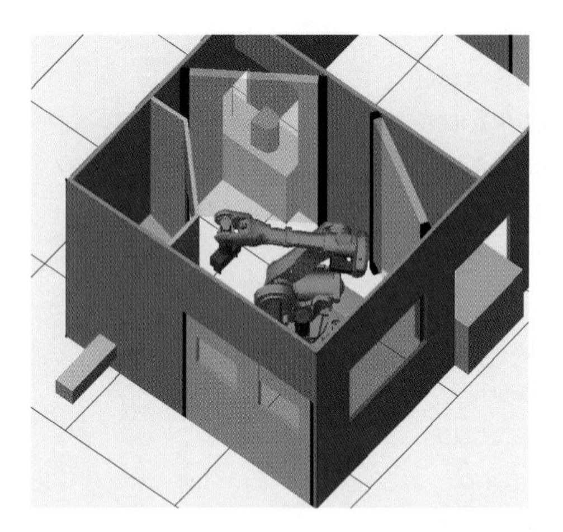

3D-planning environment for robot simulation (source: plusdrei GmbH)

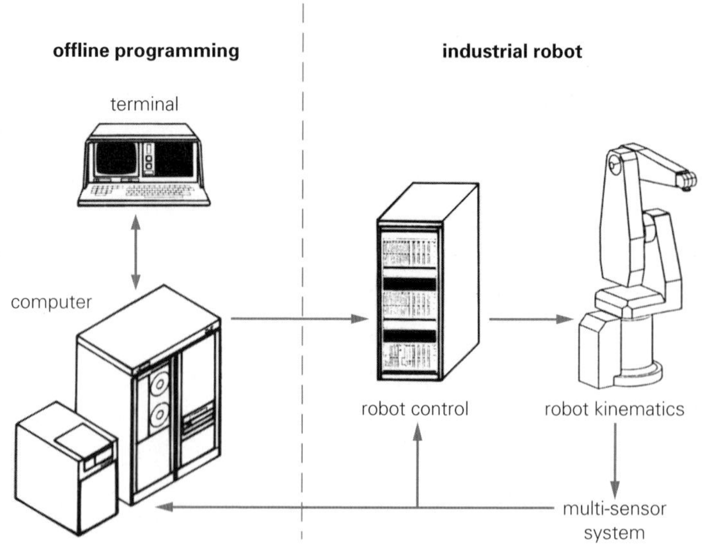

Figure 2.13 Offline programming in the early 80s

Apart from the potential stated by the VDMA which will help create further price/efficiency laps over the next few years, there are other issues related to design and the use of automation components which still are to be mastered.

The first to be mentioned in this context are planning instruments which allow for mainframe software solutions for first robot cell constructions. The efficiency of these systems and its relevance under practical aspects have been very limited so far.

Simulation efforts were not as favorable for users as necessary. Planning tools were first offered at resonable prices on the basis of PC technology. This technology has become an indispenable tool for robot construction and programming.

Simulation technology greatly profits from reduced costs for any hardware. Simulations and offline programming had to be be done with formerly expensive workstations or mainframes while today they can be realized with the help of a commercial PC.

Offline programming and simulations clearly improved in quality as far as their compliance to reality is concerned. Higher manufacturing precision, i. e. improved robot kinematics concepts, better mathematical models, and numerous application options, played an important role in technically controlled compensation of faulty robot kinematics which are caused, for example, by handling heavy weights or thermal expansion.

Simulation systems are offered, which are able to integrate the various kinematics by different manufacturers or else are supported by just one robot producer. As a rule an integrator requires the flexibility to represent robots from various producers in the simulation. Inhouse systems are not always appropriate for integrators.
They offer some advantages in particular situations with respect to the realistic representation of robot control in the simulation.

A possible trend in robot and automation component technology can be seen in the pictures of service robots venturing from factory halls into "unstructured" environments. During refueling, for example, a robot has to deal with a technical object, i. e. the car, but situations and fabricates vary. Robot fueling technology is being tested worldwide for filling up liquid hydrogen.

A system installed at the Munich airport is already under trial operation. Service robots are expected to grow in numbers, exceeding those of industrial robots. Nevertheless, the requirements are by far more diverse so that clear definitions cannot be found as easily as for industrial robots.

Due to the fact that robots move towards new applications the robot industry hopes to increase significantly the numbers of robots which are produced today. 80% of the robots today are used in automotive production. The next 10 years we will face a big shift to other industries. The robot manufacturers cop the new boundary conditions with specialized robots which are able to work for example in wet environments. Components of the automation industry like for example grippers are also highly improved in hygienic design.

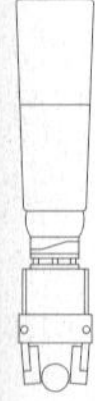

Robot fuel filling station, different car geometries have to be detected (source: Reis)

reis
Powered by Hydrogen
reis
Hydrogen

3 Getting To Grips With Handling Tasks

3 Getting To Grips With Handling Tasks

In their early childhood humans are strongly dependent on their sense of touch. Innate reflexes such as touching and gripping enable babies to experience their immediate environment. Gripping helps individuals to understand the world around them. The process of "gripping" literally includes both the act of gripping an object and understanding facts in the sense of "getting or having a grip on" them. The latter is indispensable for solving tasks. In technical language "gripping" implies more than just moving workpieces from one position to another, even though this is frequently overlooked in daily production.

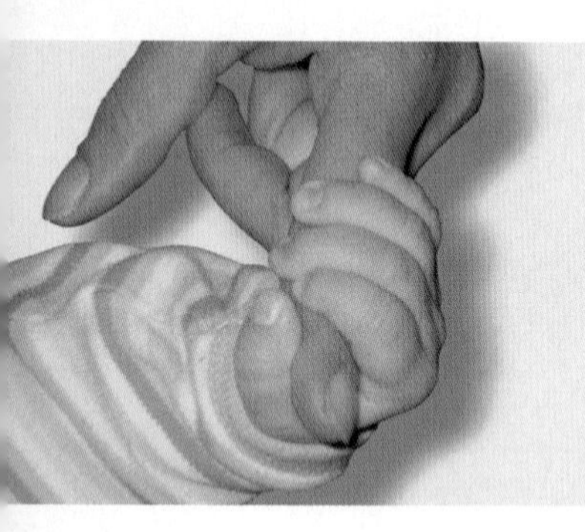

Babies use their natural gripping reflex to gain experience and get an idea of the world around them.

So far, human gripping comprises far more functions than all industrial gripper systems together. Gripper development (Chapter 2) shows that a gripper will be expected to do more than simply pick and place workpieces. Gaining information on workpieces and ambient conditions of a specific handling task becomes more and more important for process reliability and offers considerable potential for faster handling applications at a lower cost.

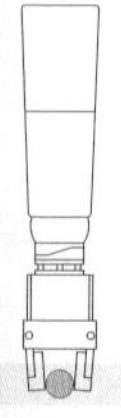

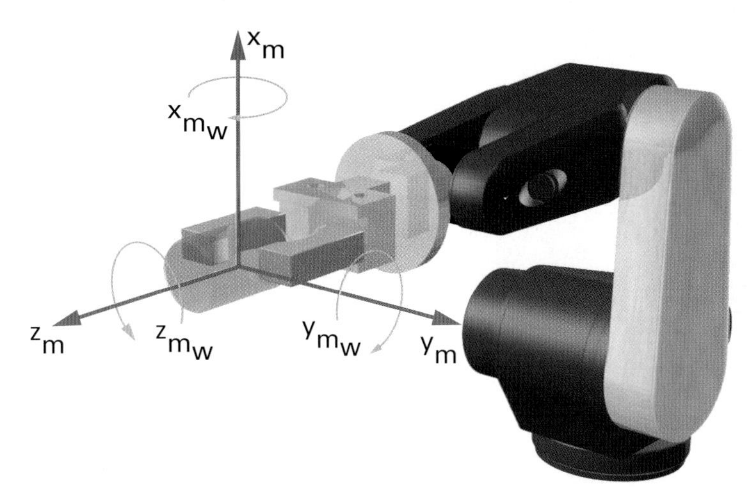

Figure 3.1 Workpiece/gripper system of coordinates, $X_m - Y_m - Z_m$

In the early 80s the VDI regulation 2860 gave a rather plain description of gripper function:

"A gripper is the subsystem of an industrial robot which maintains a limited number of geometrically defined workpieces for a set period of time, i. e. secures the position and orientation of the workpieces in relation to the tool`s or the gripper`s system of co-ordinates. This Secure function is usually built up before the moving process, maintained during the moving process, and finally reversed by releasing the workpiece."

From the current point of view this definition needs to be extended as modern gripper design and sensors offer new opportunities and we will concentrate on this reality in the following chapters.

The object of gripping, the product, the workpiece, or component, is put at the center of our initial analysis. The term "workpiece" will be used, even when there is no work being performed on the piece while it is being gripped. The workpiece can be a finished product or a product that is still being processed.

Subsequently, technical tasks of mechanical grippers for pick- and place operations and related aspects are presented and explained in detail.

Whether it is a human hand or a mechanical gripper, every gripping task is influenced by the following criteria:

- **ambient conditions of the gripping task**
- **workpiece features**
- **workpiece status at the pick operation**
- **workpiece status at the place operation**

Overall ambient conditions, e. g. opening and closing times pre-set for efficiency reasons, are a major prerequisite for selecting the appropriate gripper for a particular gripping task. They also determine which options a gripper must include, for example, if the entire system requires feedback for process reliability.

Ambient conditions may include potential risks for the handling process. Cold or damp environments as well as cleanroom operations call for customized automation technology. Ambient conditions of production, such as three-shift operation or periodical cleaning in accordance with strict hygiene requirements, will also determine either gripper construction or appropriate provisions, such as frequent maintenance intervals, to ensure high availability. Such requirements can be mastered by technical measures. More challenging are ambient conditions which are hard to detect, such as fluctuations in workpiece quality.

Based on workpiece features, which include geometrical as well as physical properties, the type or operating principle of the gripper needs to be selected.

Additional information on the workpiece status, especially the position and orientation of the workpiece within the workspace, are essential for safe gripping. Similar information must be available on the position where the workpiece is placed. Pick operation and place operation need to be treated as separate processes.
It may occur that a workpiece is perfectly picked up by the gripper, yet placed with difficulty as the gripper fingers are not able to open within a particular workspace at the place station.

ambient conditions of the gripping task	**workpiece features workpiece properties**	**workpiece status during pick- or place operations**
economic efficiency requirements	geometry	order status
safety provisions	form	set-up tolerance
installation requirements	dimension	pick- or place operations / at rest or in motion
temperature	tolerance	accessibility
energy supply	mass	
cleanroom	center of gravity	
hygiene	surface	
maintenance-free	material	
foreign substances	consistency	
workpiece variety	workpiece behavior	
monitoring required		

Table 3.1 Parameters influencing the gripping task

3.1 The Workpiece is the Starting Point

Gripping tasks are so multifaceted because of the varied geometrical and physical properties of workpieces. Moreover, the same workpiece can be "presented" to the gripper in different ways.

Before workpieces are gripped one needs to look at the characteristics which define the workpiece type and the way the gripper is supposed to pick it up. Thereafter, the workpiece status needs to be established, i. e. how the workpiece is "presented" to the gripper.

In table 3.2 workpiece features which are of major importance for the gripping process and affect the construction of grippers and their principle of function are listed. Workpiece features can be divided into workpiece characteristics and workpiece behavior.

One major feature is workpiece geometry as it defines the options for applying force to a workpiece. The characteristic form elements mainly serve for workpiece orientation and are essential for workpiece insertion. Physical workpiece characteristics determine the way forces are applied. The latter is of particular interest for handling delicate workpieces since the surface cannot be damaged.

Workpiece:
rotor for an electromotor

Workpiece behavior is defined when the workpiece is at rest and when it is in motion. Workpiece behavior becomes most interesting before the pick operations starts. A workpiece which may roll must be kept steady to prevent it from changing position while it is about to be picked up.

In table 3.3 different workpiece geometries are defined and described in an exemplary manner.

workpiece characteristics			workpiece behavior	
workpiece geometry	**characteristic form elements**	**physical characteristics**	**at rest**	**in motion**
• form (behavior type) • extension / dimension • lateral proportions • symmetry • size	• bore • rod-stop, flange • crimp, bead • notch • rift • groove, slot • chamfer • hook • release • camber/sweep	• material • center of gravity • rigidity • ultimate strength • mass • surface • temperature • processing time • gliding property	• stability • stable orientation • preferred orientation • stackability • suspendability	• slidability • rollability • directional stability

Table 3.2 Essential workpiece features

spherical	cylindrical	cubical	plane	tubular	fungi-form
spherical workpieces and variations, e. g. ball-bearing balls	cylindrical workpieces without variations, with length-diameter ratio 0.5<l/D<30, e. g. smooth bolts, shafts, polls	block or bulky workpieces of prismatic form, e. g. cube, three-edge, four-edge	mainly two-dimensional workpieces, in most cases already in plane preferred orientation	thin- or thick-walled, not completely closed tubes of cylindrical, prismatic, conical, or mixed forms	simple workpieces of cylindrical, prismatic, or coniform geometry, e. g. screws, rivets

pyramidal / coniform	regular forms	irregular forms	entangling goods	others
workpieces of regular and general pyramidal and conical form, e. g. wedge, dual wedge, full cone, truncated cone	workpieces with more than one ledge or variations with linear axes	workpieces with linear and/or crossing axes mainly massive, e. g. press or forge slugs	workpieces bound to jam or wedge, e. g. piston rings, coil springs	all materials workable from a roll, poll, or strip, e. g. steel tape, wire, etc.

Table 3.3 Main workpiece geometries

In most cases, we as humans know exactly how to handle a workpiece when looking at its geometry. Our hands automatically adapt to the surface or form of it and permit a safe grip. In case the workpiece is about to slide through our hands, we can tighten the grip by increasing the friction forces on the surface and thus secure the workpiece. We naturally adapt acceleration and final speed of this movement to the mass of the workpiece.

The workpiece geometry allows first statements to be made on the type of force induction. Workpieces in block or cylinder form have clearly defined surfaces for force induction. Cones or irregular forms are rather difficult to grip because their surfaces are not parallel to each other. Plane workpieces are generally easy to grip but their surfaces are not alway easy to access. Irregular forms make it more difficult to define surfaces appropriate for force induction.

The way the workpiece is presented to the gripper determines if and how a force can be applied. Table 3.4 helps to roughly distinguish between single workpieces and unsorted or sorted workpiece compounds.

Table 3.5 describes the effect of workpiece geometry and physical characteristics on workpiece behavior. The behavior of a single workpiece and its compound behavior, i. e. when interacting with other workpieces, must be differentiated. It is important to find out if the workpiece needs to be isolated or if it falls into a preferred position to facilitate the gripping process. The preferred position is the final position a workpiece maintains after landing on a surface from free fall. A workpiece can have one or several preferred positions depending on its form and mass distribution.

Workpiece behavior should be defined by appropriate pre-tests as inconspicuous details such as burrs, flutes, or other fluctuations in surface quality, etc., may be essential for process reliability. Additionally, it is important to establish any limitations to force induction which may result from workpiece material or surface quality.

	spherical	**cylindrical**	**cubical**	**plane round**	**plane angular**	**tubular**	**pyramidal / coniform**	**fungi-form**	**regular forms**	**irregular forms**	**entangling goods**	**others**
single work-piece												
unsorted												
sorted												

Table 3.4 Examples of single workpieces or workpiece compounds

	spherical	**cylindrical**	**cubical**	**plane round**	**plane angular**	**tubular**	**pyramidal / coniform**	**fungi-form**	**regular forms**	**irregular forms**	**entangling goods**	**others**
at rest												
falling (open)												
sliding												
rolling												

Table 3.5 Examples of workpiece behavior in relation to workpiece geometry

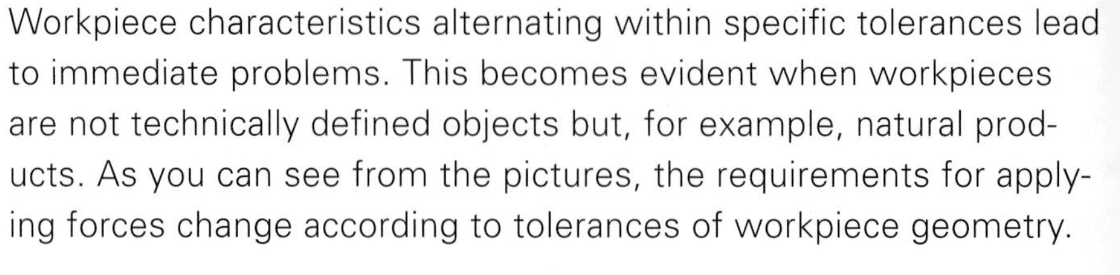

Workpiece characteristics alternating within specific tolerances lead to immediate problems. This becomes evident when workpieces are not technically defined objects but, for example, natural products. As you can see from the pictures, the requirements for applying forces change according to tolerances of workpiece geometry.

Apples as an example of differing measuring tolerances

Castings as an example of differing measuring tolerances

Measuring tolerances, which are common for natural products, are not the rule in the metal manufacturing industry but occasionally do occur. Workpieces of the same type may differ within one and the same series in terms of surface quality, such as cutting oils left on parts or filing residues after workpiece processing. In this respect the metal-working industry as well as the food processing industry are faced with changing gripping conditions depending on the workpiece. Therefore, the forces applied on differing workpiece surfaces may change from workpiece to workpiece and require safe calculation.

As shown in table 3.5 position and orientation of the workpiece are important to adjust the gripper position accordingly. The position of a workpiece before the pick operation already sets limits to force induction options as the points where forces can be applied may not always be accessible.

A workpiece has translatory and rotary orientation options or so-called degrees of freedom (dof). These are described in a system of co-ordinates to define the translatory and rotary degrees of freedom.

The so-called order status of the workpiece is defined to describe the situation of the workpiece in relation to its position and rotation in the workspace. The workpiece`s translatory and rotary degrees of freedom are used for this purpose, too.

Two degrees of freedom are distinguished:

orientating degree (OD) = rotary degrees of freedom

positioning degree (PD) = translatory degrees of freedom

The table lists all different orientation and positioning degrees of a workpiece. The respective orientation and positioning degrees are rated from 0 to 3. Rating 0 stands for an entirely undefined orientation or positioning degree, the overall rating is counted in even numbers up to rating 3 for an exactly defined position.

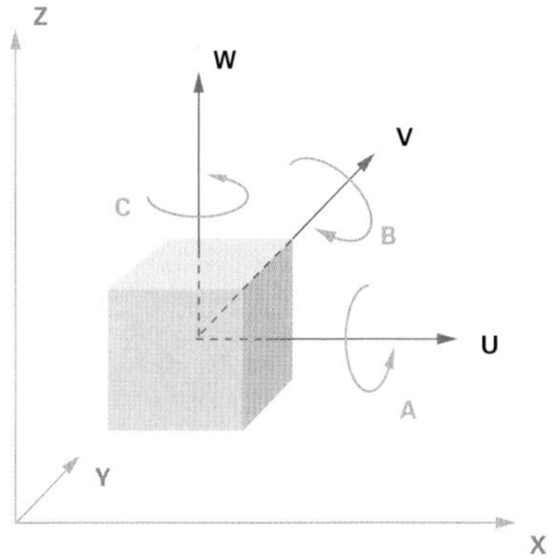

Z
W
V
C
B
U
A
Y
X

XYZ reference system of coordinates
UVW translatory degrees of freedom
ABC rotary degrees of freedom

Figure 3.2 Translatory and rotatory degrees of freedom

degree of orientation (OD)		positioning degree (PG)	
3	orientation of the workpiece defined for all rotary axes	3	source of the workpieces`s own system of coordinates in one defined point
2	orientation of the workpiece defined for two rotary axes	2	source of the workpieces`s own system of coordinates optionally on a curve (e. g. straight line or circular path)
1	orientation of the workpiece defined for one rotary axis	1	workpiece situated on a surface (e. g. plane, cylinder case)
0	orientation of the workpiece not defined for any rotary axes	0	workpiece situated anywhere in the workspace

Table 3.6 Definition of workpiece degrees of orientation and position

Both degree of orientation and degree of position define the so-called order status of a workpiece as follows:

order status OS = OD/PD
Example: 1/0 a) rotary degrees of freedom (dof) defined b) translatory degrees of freedom (dof) not defined

	3-dof not defined	1-dof defined	2-dof defined	3-dof defined
bulk goods	0/0	1/0	2/0	3/0
plane distribution	0/1	1/1	2/1	3/1
linear distribution	0/2	1/2	2/2	3/2
isolated	0/3	1/3	2/3	3/3

Table 3.7 Order status of a workpiece (dof: degrees of freedom)

The order status determines the effort necessary for a workpiece to be picked up or to be moved around a defined number of axes by the gripper before the handling process can be considered complete. Practical experience shows that most differing tasks may occur which increase the order status.

Table 3.7 lists all order status options of a workpiece.

Changing the order status of workpieces is the core of handling tasks, regardless of whether unsorted workpieces are stored in a transport box to be fed one by one into a processing machine, or whether concentric gripping is sufficient for further processing. A pick operation can generally be assumed to be more difficult in case the workpiece is on the lower end of the order. This frequently occurs during handling tasks for which the translatory coordinates are changed, e. g. when workpieces are taken from one pallet and put onto another one.

Once a workpiece has reached a certain order status it is generally recommended to maintain this order status. This seems self-evident but cannot always be realized in practice due to specific workspace conditions or manufacturing processes.

Workpieces tend to loose their order status when being transported on several conveyor belts. Designing conveyors or restructuring whole transport systems under the aspect of maintaining the workpiece order status is laborious and costly.

Storage as bulk goods (above) and final clamping position (below)

Workpieces are frequently processed as bulk goods for production and thus change from a sorted situation into an unsorted one.
If interim storing is indispensable, storage space must be used effectively to keep inventory costs low. Storing workpieces in orientation, e. g. on pallets, is very expensive in most cases because each workpiece geometry would require specially fabricated pallets.

Considering increasing workpiece variety and decreasing lot numbers this puts producers to unreasonable expense. Therefore, workpieces in need of interim storing are frequently stored as bulk goods. The pictures show both pre-manufactured parts stored as bulk goods and workpieces in their final clamping position.

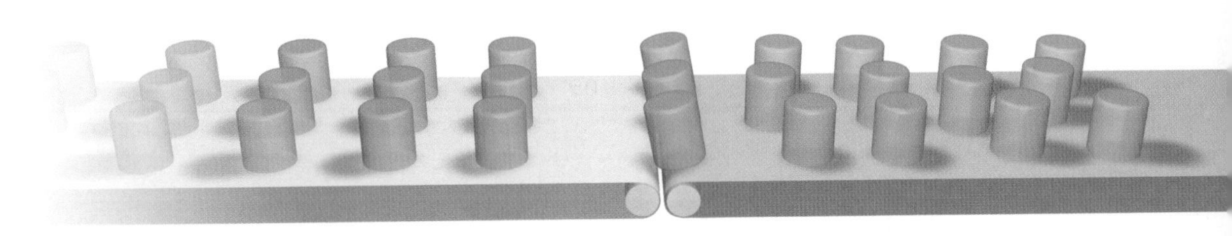

Changing order status of workpieces during transport on different conveyors

order status options for workpieces

a) cube, no symmetrical axis

<< workpiece position / workpiece orientation >>			
0.0			
1.0			
2.0	2.1	2.2	
		3.2	3.3

workpiece status	matrix-field
1. bulk goods	0.0
2. plane	1.0
3. linear	2.0
4. rotate around v	2.1
5. rotate around w	2.2
6. isolated	3.2
7. rotate around u	3.3

b) disk, rotary symmetry around w

<< workpiece position / workpiece orientation >>			
0.0			
	1.1		
	2.1		
	3.1		

workpiece status	matrix-field
1. bulk goods	0.0
2. fplane, at parallel rotation around u or v (for preferred orientation)	1.1
3. linear	2.1
4. isolated	3.1

Table 3.8 Effect of workpiece symmetries on the order status

According to workpiece geometry, an order status is more or less difficult to establish or to maintain. Table 3.8 shows the differences between cube and disk forms. An order status 3/1 for disk forms already describes the position of the workpiece in full because of its rotary symmetry around the w axis.

Generally it must be observed in how far workpiece symmetry is dependent on more than its geometry. If workpieces differ in surface features as a result of changing their position they are to be treated like assymetrical parts.

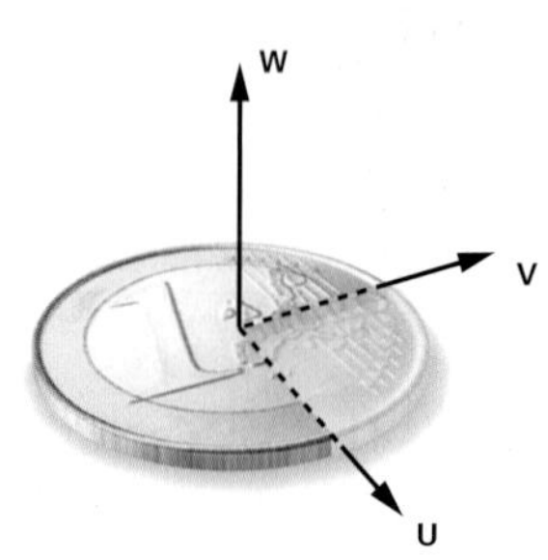

An example of this includes coins which are put into a collector`s booklet. The size of the coin is just as important as the difference between front (averse) and back (reverse) of the coin. The coins not only need to be inserted front-up but aligned accordingly to achieve a representative look as explained in the picture.

For each task or handling situation it is important to describe the order status of the workpiece before and after handling. From this description we can see how many axes or degrees of freedom the handling unit must provide in order to completely solve the task.

The unlimited variety of workpiece features demands high gripping flexibility. The number of workpiece features a gripper copes with corresponds to the number and type of force induction options. The more options to apply force onto a workpiece a gripper offers, the more workpiece geometries can be gripped. The operating elements or gripper fingers of a gripper can be designed or adapted accordingly.

3.2 Gripper Fingers As Operating Elements

Forces are transmitted by gripper fingers, the so-called operating elements of the gripper. The amount of force which needs to be applied depends on the body mass, surface friction, and geometry of the workpiece. Workpiece geometry defines criteria such as:

- **distance between force induction point and mass center of gravity**
- **mass moment of inertia**
- **type of force induction**

If gripping force just needs to be transmitted via surface friction, pressure must be put on the workpiece surface. For workpieces which easily react to pressure, e. g. the surface of which is easily deformed or damaged, a maximum pressure must be determined. For safety reasons maximum pressure during gripping must be clearly lower than the approved pressure for the respective material. Calculations on maximum pressure for different contact bodies are shown in table 3.9 distinguishing point and linear contact between gripper fingers and workpieces.

Gripping forces vary according to form and number of active surfaces between workpiece and gripper fingers. In table 3.10 three typical combinations of force-fit gripping are compared. The influence of surface types on gripping force is expressed by the respective formule.

Differing coefficients of adhesive friction for defined material combinations are detailed in table 3.11.

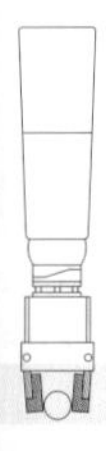

	case	contact bodies		size radius R	deformation and max. pressure
point contact	1		sphere	$R = r_2$	$p_{max} = 1{,}5 \frac{F}{a^2 \pi}$
	2		hollow sphere	$R = r_2$	$a = \sqrt[3]{\frac{1{,}5(1-v^2)Fr}{E}}$
	3		plane	$R = \infty$	$\frac{1}{r} = \frac{1}{r_1} + \frac{1}{R}$; $E = \frac{2E_1E_2}{E_1+E_2}$
linear contact	4		cylinder	case 1	
	5		holllow cylinder	case 2	$p_{max} = \frac{2F}{\pi bl}$
	6		plane	case 3	$b = \sqrt{\frac{8Fr(1-v^2)}{\pi El}}$

Table 3.9 Maximum Hertz pressure

	gripping force	basic gripping force	influencing factor of contact type	corrective factor for contact safety
	$F_G =$	$F_R \cdot \frac{S}{\mu}$	$\frac{1}{2}$	
	$F_G =$	$F_R \cdot \frac{S}{\mu}$	$\frac{\cos\alpha}{2}$	k
	$F_G =$	$F_R \cdot \frac{S}{\mu}$	$\frac{\cos\alpha}{1+\cos\alpha}$	

Table 3.10 Gripping force calculation for various finger forms

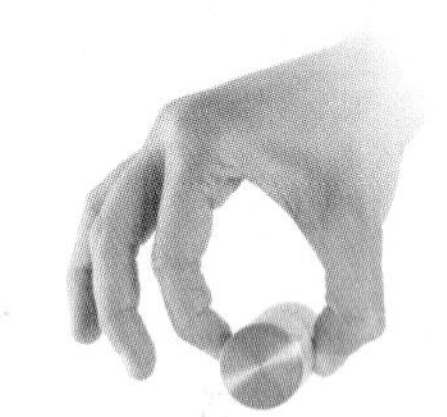

material combinations	coefficients of adhesive friction	
	dry	lubricated
Steel on cast iron	0.2	0.15
Steel on steel	0.2	0.1
Steel on Cu-Sn alloy	0.2	0.1
Steel on Po-Sn alloy	0.15	0.1
Steel on polyarid	0.3	0.15
Steel on friction coating	0.6	0.3
Steel on Quentes (SCHUNK)	0.3 - 0.4	-

Table 3.11 Coefficients of adhesive friction for different surface material combinations

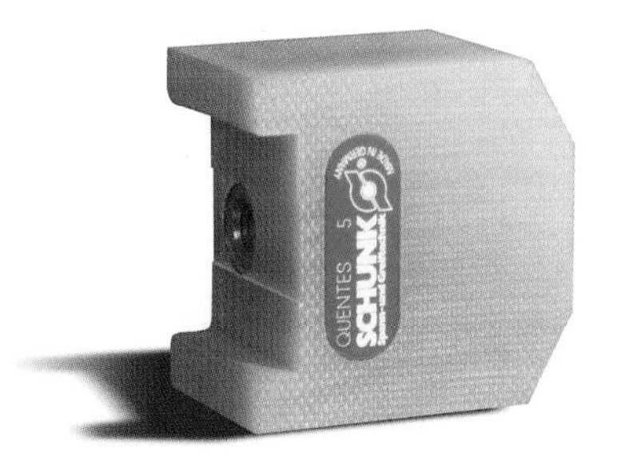

Special gripper fingers made from Quentes

Coefficients of adhesive friction are stated for dry and lubricated state of the surfaces. Table 3.11 illustrates that the amount of gripping force depends on the friction qualities of both workpiece and gripper surfaces. Gripper producers offer materials which combine good friction quality with high stability.

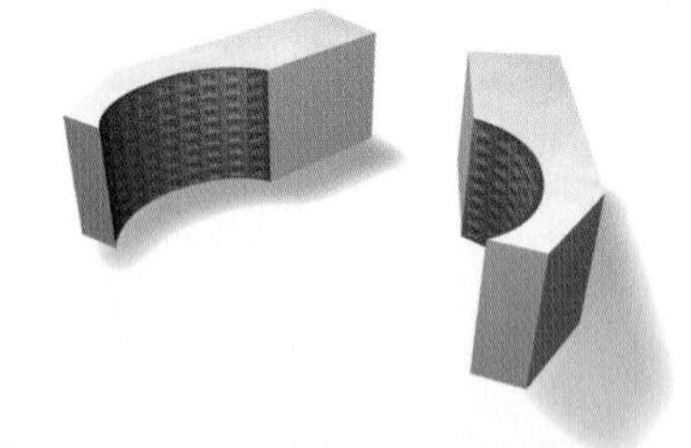

Good contact between workpiece and gripper fingers is essential for safe gripping at minimum force. Therefore, it makes sense to have maximum surface contact between the workpiece surface and the operating elements of the gripper. A special coating (compare table 3.11) can further reduce the forces required. So-called adhesive cushions made of elastomers combined with an aluminum support plate are available. Elastomers offer particularly good friction while the aluminum support plate ensures stability.

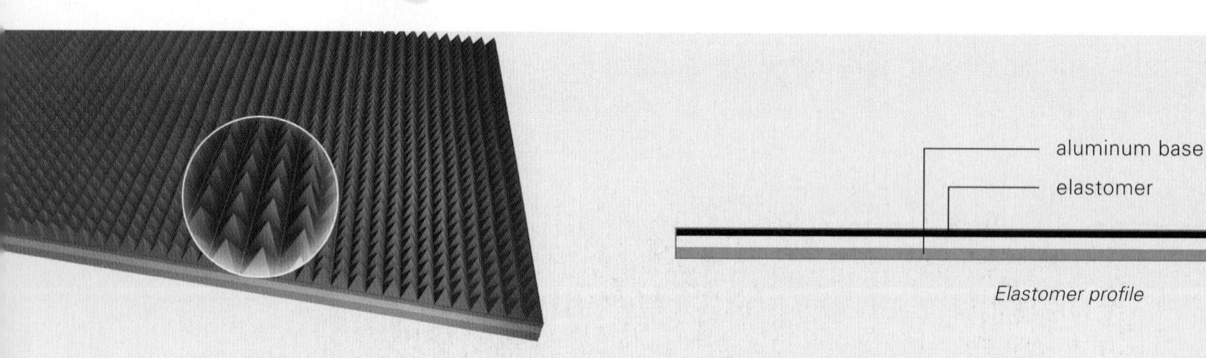

Elastomer profile

Elastomer plan view

Besides the ideal combination of workpiece and gripper materials friction can be increased by form-fit gripping which is achieved by a suitable profile such as scoremarks or teeth. This type of gripping force reduction is perfect for handling workpieces without surface restrictions. Cast iron, for example, can be gripped with hard metal chucks which guarantee form-fit gripping. These carbide chucks can handle loads up to 30,000N and are simply replaced after wear-out.

Form-fit gripping is usually applied to workpieces which are due for further processing.

With the help of these methods gripping force can be distributed better on the workpiece`s force induction areas. Distributing forces by means of maximizing these areas is one option to reduce gripping force. Table 3.12 illustrates different force induction options for spherical workpieces. The best solution is to enclose workpieces with either two or three gripper jaws which are adapted to workpiece radius.

For the respective handling task either the gripper finger or the force induction area of the workpiece need to be designed for optimum results.

Types of movement (translation or rotation) of the operating elements may vary according to gripper construction. As illustrated, workpieces of different dimensions may not offer identical contact surfaces which may lead to inaccurate positioning of workpieces within the gripper. Spherical workpieces are especially suitable for using one gripper jaw design for various workpiece dimensions. Nevertheless, it may lead to critical problems if the gripper fails to place the workpieces in one and the same position (movements of gripper jaws). Precautions have to be taken to prevent feeding the gripper system with the wrong type of workpiece as this is bound to provoke collisions during the pick operation.

The form of the workpiece sets limitations on the force induction options through the gripper fingers. As a result, careful gripper design and construction is essential for safe gripping. Force-fit gripping as well as form-fit gripping are options for securing the workpiece within the gripper according to the respective form and moving task. Form-fit and force-fit gripping are frequently combined.

Table 3.12 gives a structured overview on the gripper options for pick operations according to number of contact surfaces and type of gripping. Force-fit gripping is also called friction lock or force lock, form-fit gripping is also known as form lock. This overview shows that human gripping includes all gripping types except for adhesive gripping (neglecting sticky fingers, of course), offering very high flexibility for the gripping process.

	principle of function		force lock	form lock
gripping with one contact surface	adhesive grip		●	
	reverse grip			●
gripping with more than one contact surface	force-fit		●	
	force-fit/form-fit		●	●
	form-fit			●

Table 3.12 Classification of gripper principles of function according to form-fit and force-fit options

hard metal insert for grippers

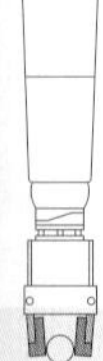

	rule	not appropriate	appropriate
1	use form-fit before force-fit gripping		
2	use anti-sliding measures for adhesive grippers		
3	provide parallel gripping surfaces		
4	design gripping surfaces with due diligence		
5	put center of gravity into the center of gripper fingers		
6	use the same gripping points for different workpieces		
7	avoid gripping sensitive surfaces (p = polished)		p
8	prop thin-walled workpieces and restrict gripping force		
9	create plane adhesive surfaces		
10	provide centering elements for high-precision gripping		

Table 3.13 Rules for workpiece design appropriate for gripping

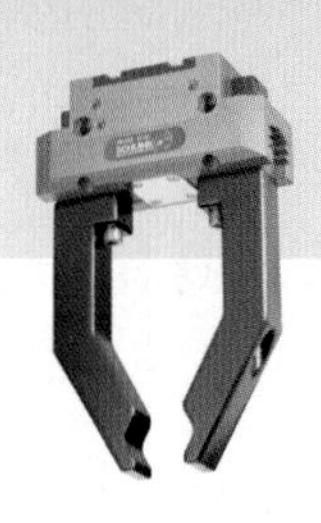

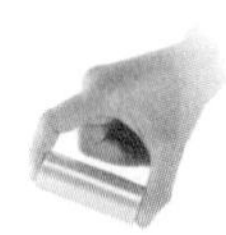

Gripping with one contact surface, the so-called adhesive grip, is the classic operating principle for suction grippers, magnetic grippers, or adhesive grippers. The reverse grip is sometimes used for extremely difficult gripping tasks such as coping with very unstable workpieces. The classic reverse grip merely holds the workpiece by the principle of gravity.

Form-fit and force-fit gripping are basic principles of gripping with more than one contact surface. Combinations of form-fit and force-fit gripping are frequently used. As explained earlier, the human hand is superior to any technical systems in terms of gripping flexibility as it naturally combines numerous gripping types.
A presumably simple gripping process, such as picking up a coffee-cup, comprises most diverse gripping principles and several gripping qualities.

The users of gripping technology must usually rely on their experience with the relevant workpieces in order to help design gripper fingers or gripper jaws. Gripper producers provide the user with a great variety of standard gripper jaws which are available as accessories for basic grippers.

The basic gripper defines the interface for gripper kinematics to ensure compliance in all respects. Finger blanks are easy to assemble, equipped with the respective clamping contours, and available in aluminum, steel, or synthetic versions. In terms of designing operating elements the workpiece-related and the gripper-related figures are important. When combining contact surface qualities the workpiece-related figures, e .g the form of the workpiece, have direct influence on the gripper-related figures, e. g. the contact surfaces.

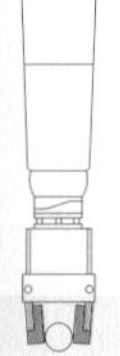

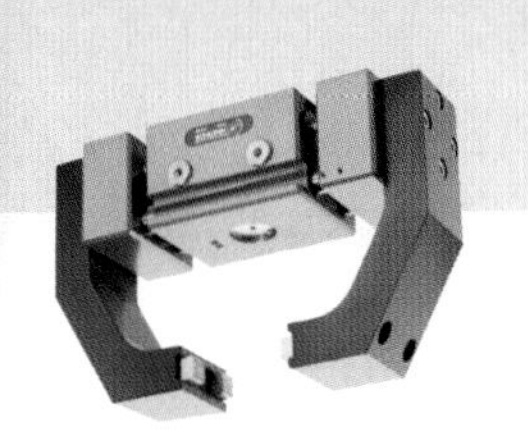 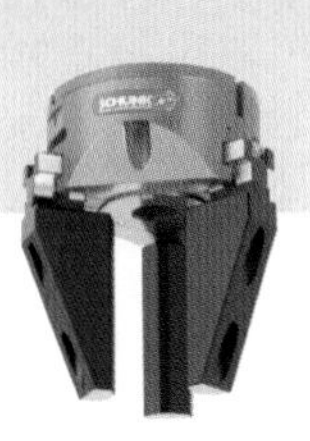  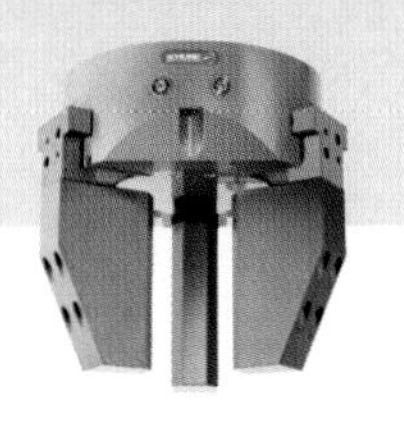 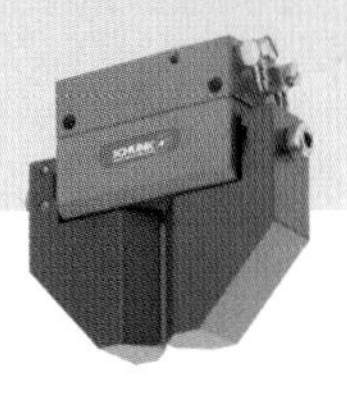 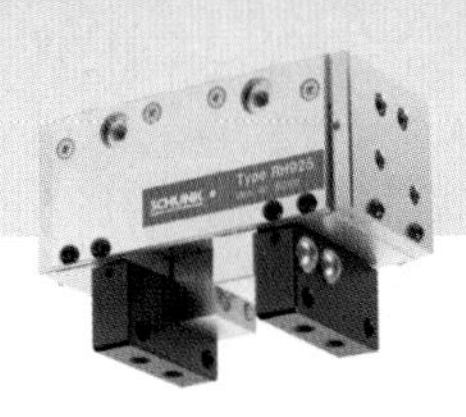

range of gripper fingers

The design of operating elements always depends on a specific task such as piston rod handling. The piston rods need to be transported in suspension and inserted into the processing unit.
The gripper in the picture copes with eight piston rods at a time to increase the cycle time of the handling process. The piston rod place operation is determined by the processing unit and requires very sensitive gripper fingers. The principle of gripping is friction lock, and high gripping force is required for process reliability.
As the fingers are both slim and long, strong gripping force causes substantial tension within them. With the help of the Finite-Element-Method (FEM) forces can be visualized in modern workshop places equipped with CAD (Computer Aided Design) applications. Visualizing forces means being able to monitor them in order to avoid the risk of gripper fingers being deformed.

stationary contact surfaces			**flexible contact surfaces**		
single surface	several separate surfaces	several unseparated surfaces	deformable surfaces	movable surfaces	switch-off/switch-on surfaces
• round jaws • prism jaws • comb jaws	• dual round jaws • any combination of forms	• gripper jaw with inserted jaw shells	• granulate • magnetic powder • elastic finger • plasticine	• sliding surfaces • adjustable surfaces	• adhesive jaws • electromagnetic or permanent magnetic jaws

Table 3.14 Options for contact surface design

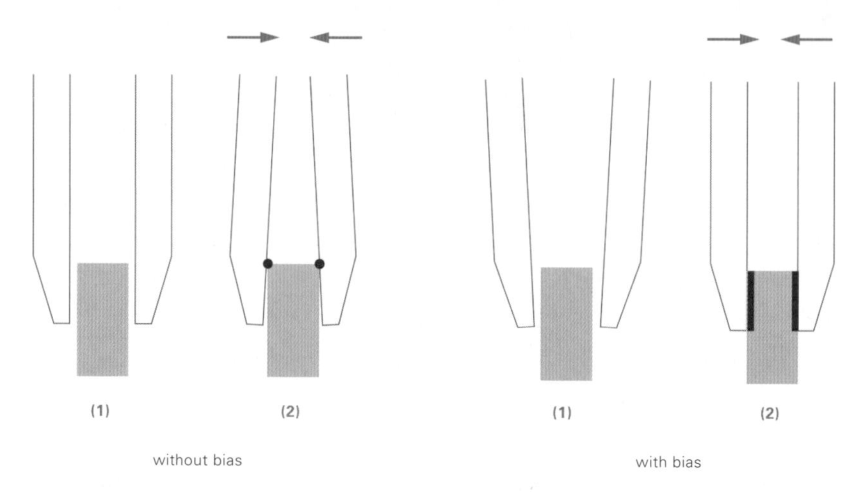

Figure 3.3 Avoiding point contact with the workpiece by gripper jaw bias

The special construction in this example is necessary because high gripping forces and long fingers may extend the gripper fingers which again may lead to unwanted point contact.

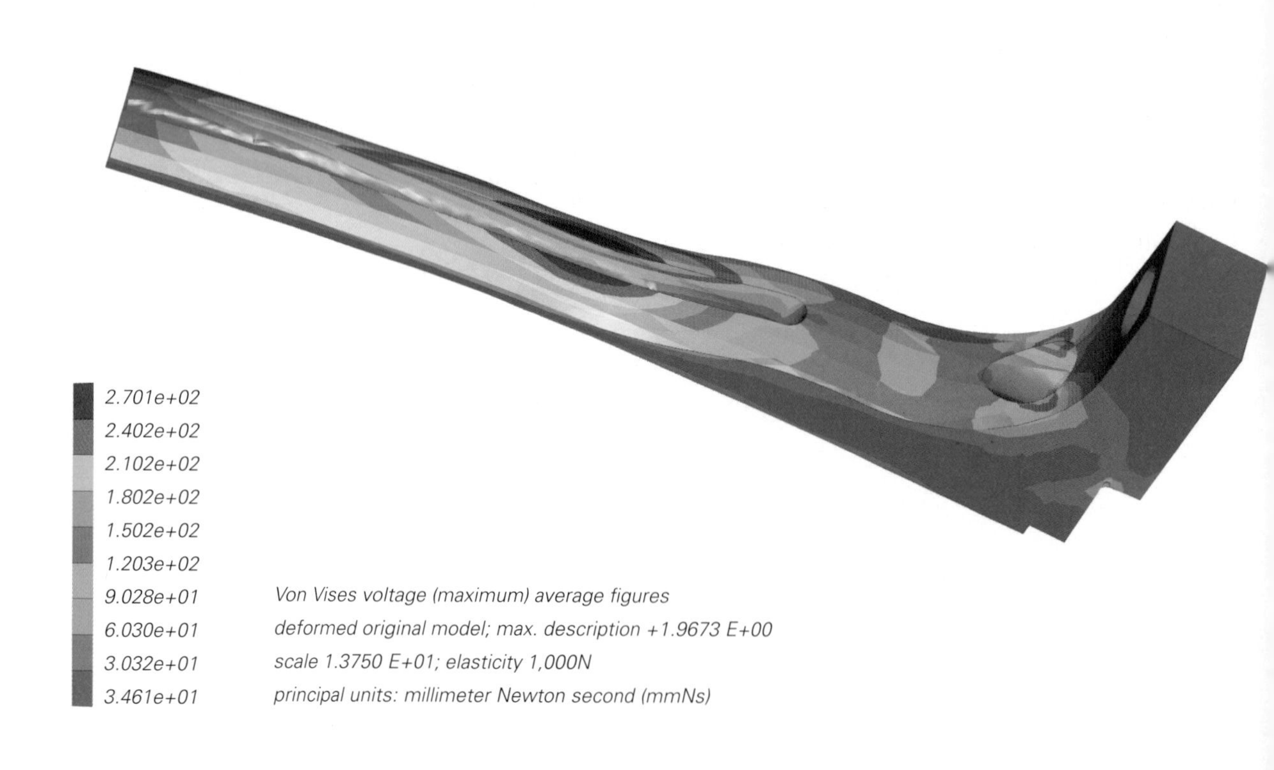

In case there is just one contact point between the workpiece and the gripper jaws, the piston rod may swing during acceleration or slow-down, which must be excluded to achieve accurate positioning. Precise calculation of the bias in the gripper fingers helps to avoid point contact. The bias ensures that gripper fingers cannot close in parallel position without load. If the fingers close around the workpiece, gripper and workpiece have the desired surface contact which prevents the workpiece from swinging at random.

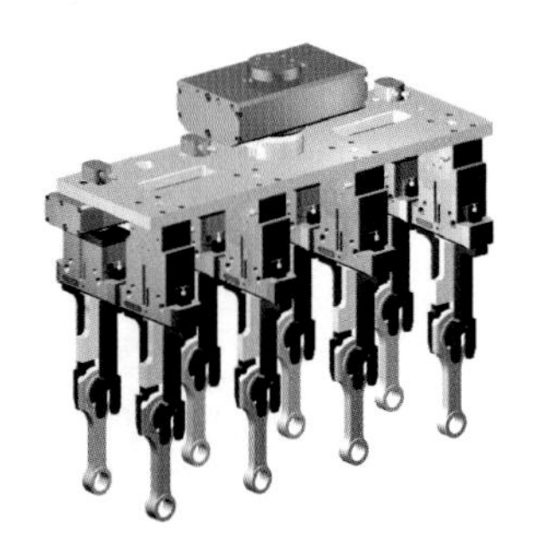

octuple gripper

Contact surface design offers two completely different options. Stationary contact surface types for gripper fingers is one of them. An alternative for highly flexible grippers are flexible contact surface types. In table 3.14 both options for contact surface design are detailed.

single connecting rod

Stationary contact surface types just have one pre-defined force transmission option. Separate surface types within the same gripper finger permit one gripper to pick up various workpiece diameters. In this case, however, the high bending effect caused by the longer gripper fingers needs to be taken into account.

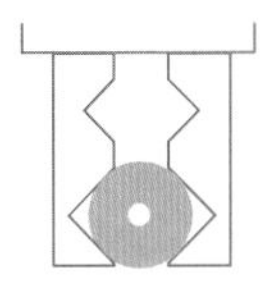

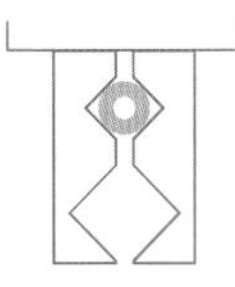

separate contact surfaces within one gripper

The better option is trying to pick up different diameters with unseparated contact surface types to keep the gripper fingers as short as possible. The number of differing workpieces is limited for this option. Some examples of gripper fingers are included to illustrate the variety of finger types and their adaptability to individual workpiece forms.

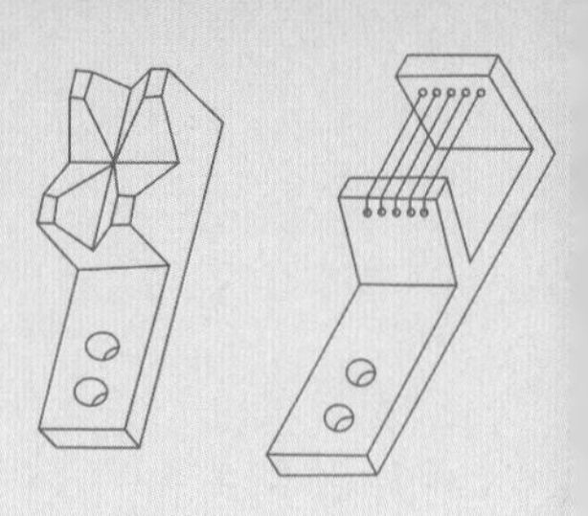

needle gripper

Gripper fingers with flexible contact surface types can handle the most diverse workpiece geometries. For example, flexible contact surfaces made of granulate or magnetic powder perfectly adapt to the form of a workpiece before they are hardened by electric energy for form-fit gripping.

A purely mechanical principle of this gripping type is a gripper equipped with densely packed nails which hold the workpiece by having it pressed into the flexible surface of nails.

Additionally, the nails can be set against each other to achieve form-fit gripping. This process is simply reversed for releasing the workpiece.

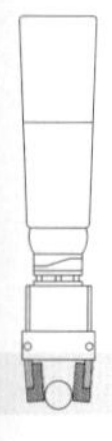

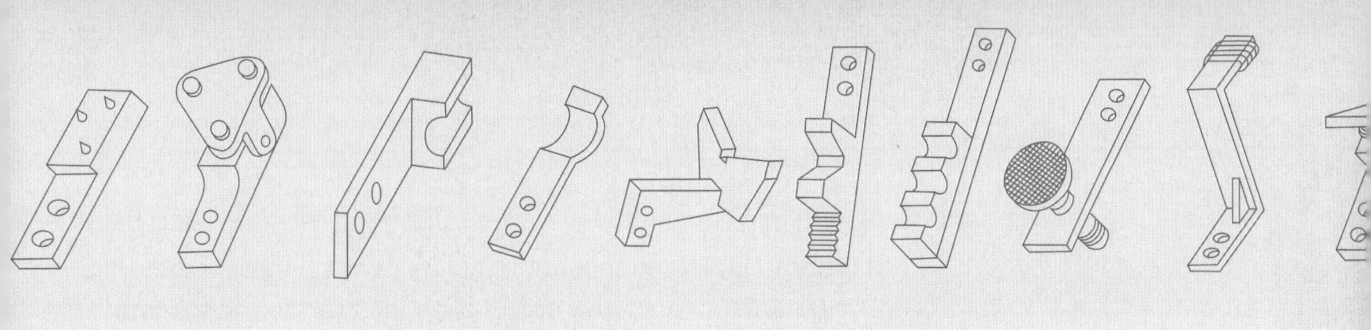

Examples of gripper jaws

The picture shows how workpieces of most diverse geometry are picked up with a most flexible contact surface. Basically it is a multi-finger gripper system, i. e. 1,200 nails have been set on 70cm^2 in order to achieve the necessary friction lock.

This gripper type, however, is suitable for robust workpieces only while delicate surfaces or workpieces would be easily damaged.

The majority of grippers which are based on the on-off switch type are so-called suction grippers. This gripping technology is a completely independent matter and not included as it would exceed the volume of this book (see definiton of contents in Chapter 1).

Suction gripper

In order to define options for gripping a workpiece it is important how the operating elements are aligned in relation to the work-piece. Three types of gripping options are distinguished:

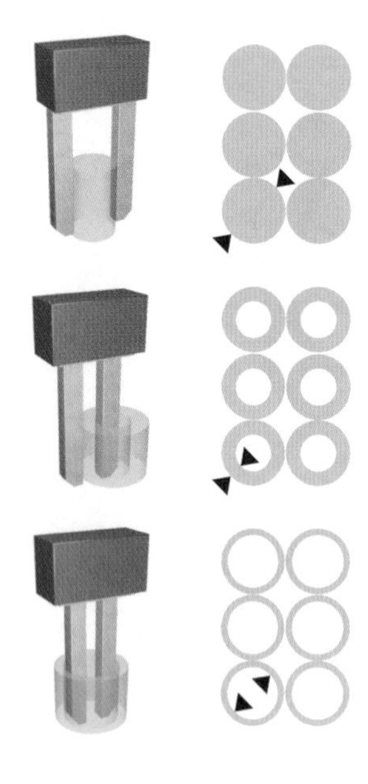

Figure 3.4 Range of gripping options

- **external grip** (gripping force is applied to external surfaces of a work-piece)
- **intermediate grip** (a workpiece, e. g. a tubular one, is gripped on external and internal surfaces)
- **internal grip** (gripping force is applied to internal surfaces of a workpiece, e. g. a bore or hole)

The benefits and drawbacks of the respective gripping option are quite evident in figure 3.4. Compact workpieces require the external grip if frictional forces are used. There is no way to pick up the round bar with any other grip. The external grip becomes problematic when the case pallet sizes are too small for pick- or place operations.

The intermediate grip as well as the internal grip can be applied to workpieces with an appropriate bore or hole. Both methods of gripping are suitable for tightly packed pallets. The internal grip is mostly used together with lathe parts because in combination with 3-finger grippers the workpiece can be well centered.

This ensures accurate positioning for releasing the workpiece no matter if contact surfaces are stationary or flexible. Flexible contact surfaces, of course, increase the number of application options.

2-finger gripper

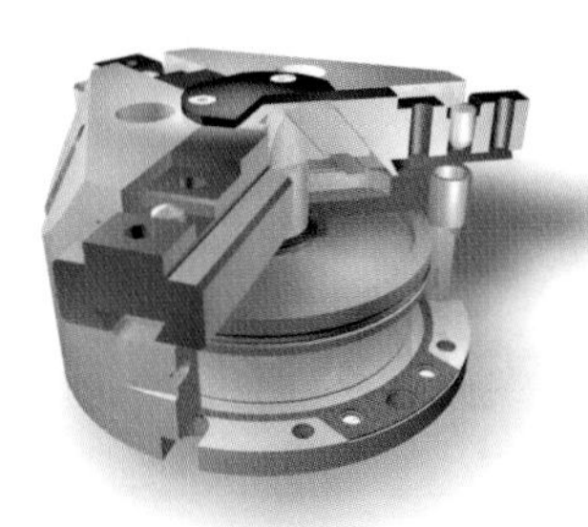

3-finger gripper

4-finger gripper

Another gripper design concentrates on the method of flexibly aligning gripper fingers in relation to the workpiece geometry. The Fraunhofer Institute for Manufacturing Engineering and Automation IPA developed an artificial hand on the model of human gripping. The thumb is situated opposite the other two fingers, and its movements are only half the size of the other finger movements.

An analysis of the human hand and more specifically the human gripping options demonstrates that only three types of grips are essential for covering a broad range of workpieces:

- **2-finger grip**
- **3-finger parallel grip**
- **3-finger concentric grip**

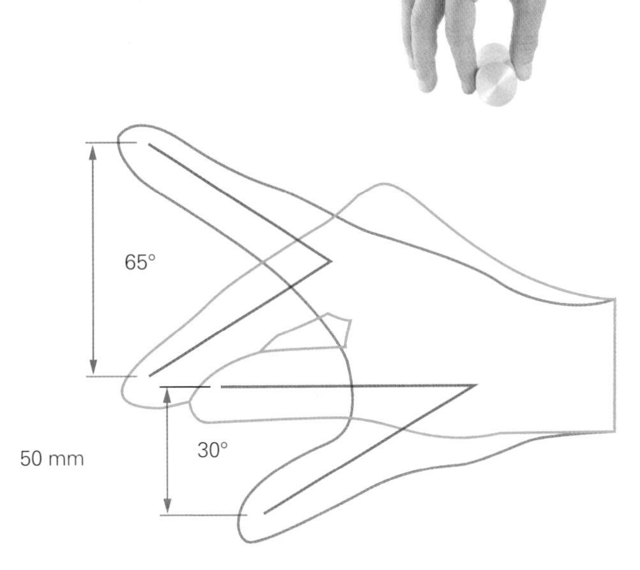

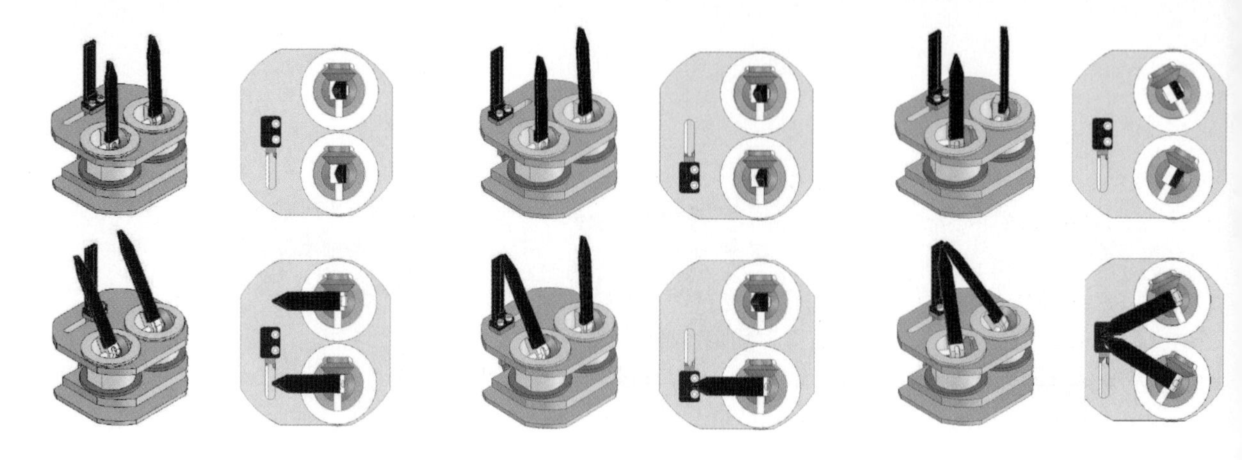

Figure 3.5 IPA Hand gripping options

These findings led to the construction of the IPA Hand which operated with one stationary gripper finger as the "thumb" and two flexible fingers. A gripper concept was developed which is capable of performing gripping types with just two electric drives as illustrated in figure 3.5. The thumb is designed to move between two stationary positions but is rigidly aligned in relation to the direction of gripping force.

Each finger can be freely programmed and runs rotating around its high axle. A two-level drive allows the gripper fingers to independently operate in three different configurations.

As contact surface alignment according to workpiece requirements is flexible, this gripper is able to cover various workpiece geometries. The gripper fingers are currently fabricated as rigid operating elements.

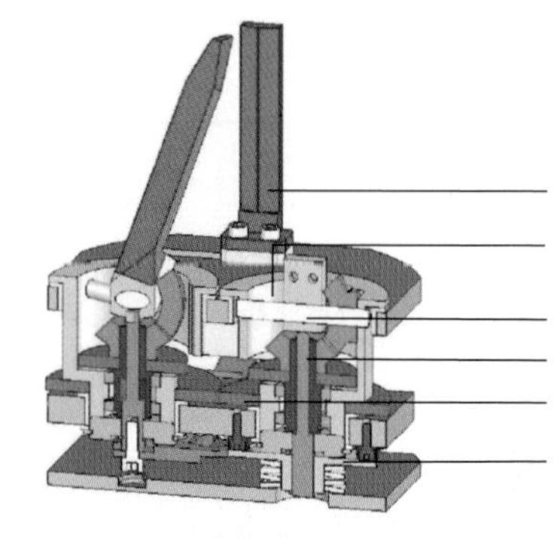

finger
housing for finger drives
excenter for coordination of finger movement
spur gears for coordination of finger positions
bevel gears for force transmission in any position
stop disc

With respect to an economically efficient operation, the following basic approaches followed:

Lower expenditure for control technology leads to the development of completely "autonomous and intelligent" components which are easy to integrate into automation systems. The overall reduction of component parts, e. g. expensive actuators, makes it possible to build compact systems of good value.

IPA Hand

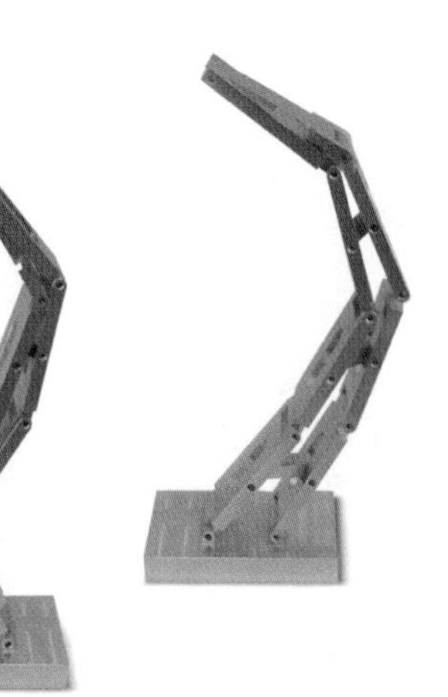

Flexible contact surface types can either be designed to be passive or active. Auto-adaptive, passive fingers kinematics comprise of single finger links which are connected by joints. The finger links contain a traverse for adapting them to the workpiece geometry when force is applied.

In case the surface contact types are actively adapted to the work-pieces the joints must be driven independently. The finger modules can be designed either as stationary elements or as flexible-passive or flexible-active elements as illustrated.

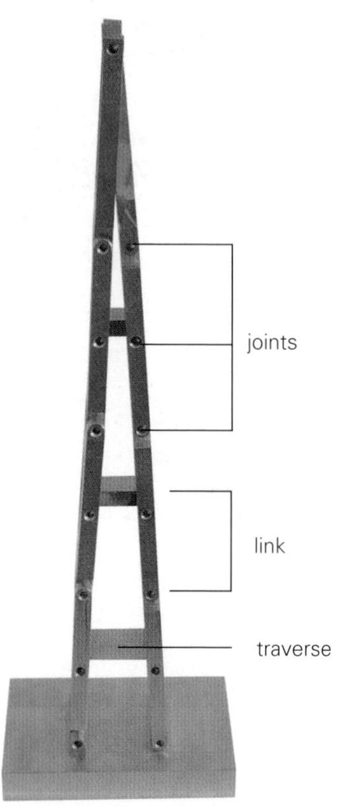

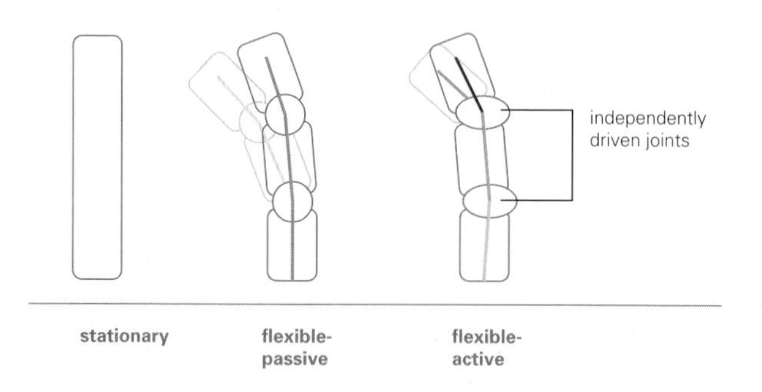

The most flexible finger concept is the flexible-active one. Many research series are being dedicated to this gripper finger concept which is based on the human model (see Chapter 2.2).

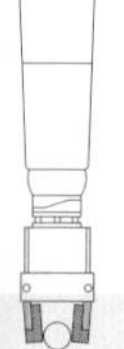

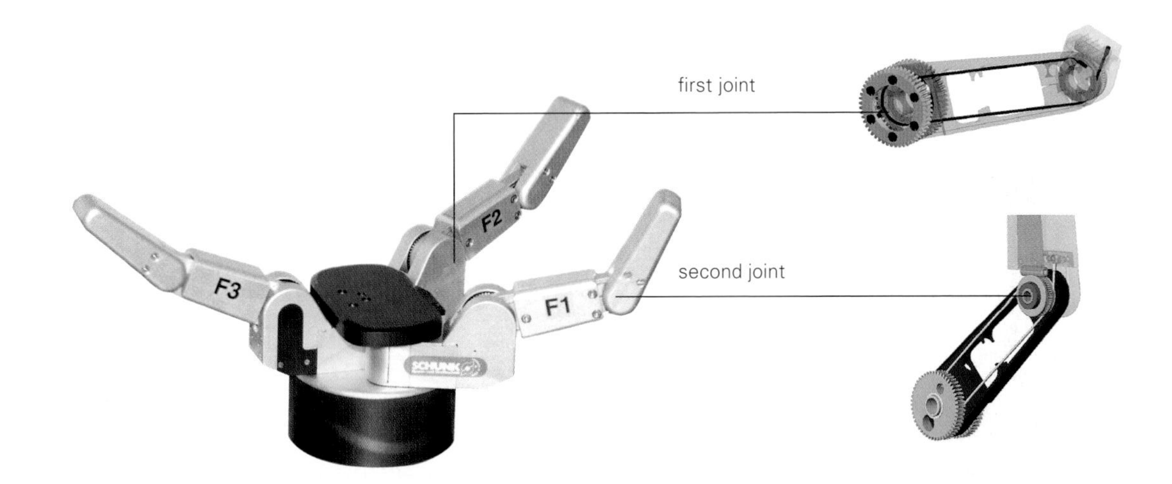

The Barret Hand is a good example for an industrially employed concept. The energy for the drive is transmitted to the finger links via worm gear units and bowden wires. The hand comprises an overall three operating elements, i. e. three fingers. Four servo-motors with distance measuring systems provide the driving energy and actuate all seven joints.

The drive of the fingers is separated so that one motor drives one finger with two joints.

Additionally the gripping force and gripping velocity are regulated. A maximum gripping force of 5N per gripper finger is achieved. The gripper weighs 1.18kg and can handle a payload up to 6kg if form-fit gripping is possible.

Adhesive or magnetic fingers can also be used to increase force on the workpiece. This force is switched on during pick operations and switched off during place operations. Therefore, the surface contact types are called switch-off or switch-on.

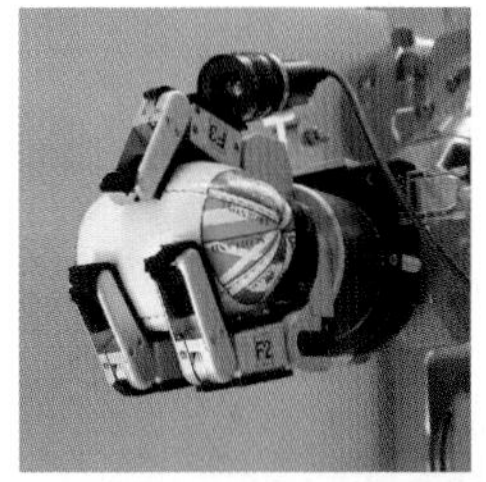

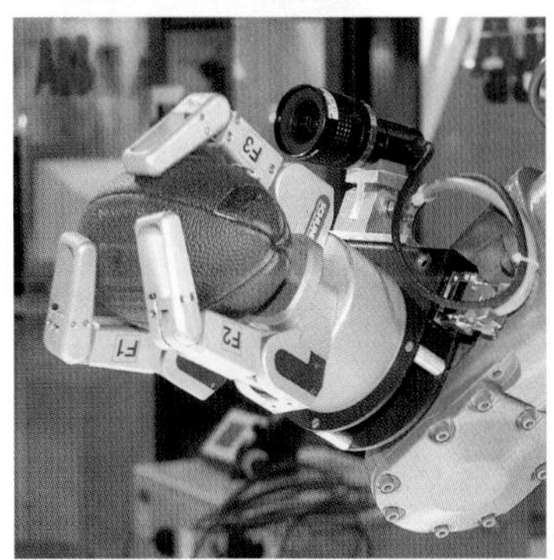

Barret Hand

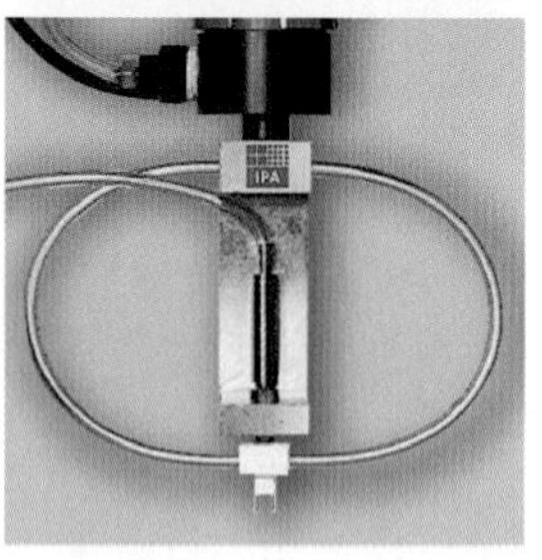

micro-components pick operation with adhesive grippers

Grippers with switch-off/switch-on contact surfaces are frequently being used in the form of suction grippers or magnetic grippers. These principles of function are easily integrated into single operating elements of grippers. The adhesive contact surface type is still a special force transmission type.

Adhesive gripper technology is based on a physical principle which is based on the surface tension of liquids. Adhesion means joining together different bodies or, in this case, workpieces, made of different materials. The cause of adhesion is molecular magnetism, the so-called adhesive forces, between the contact surfaces.

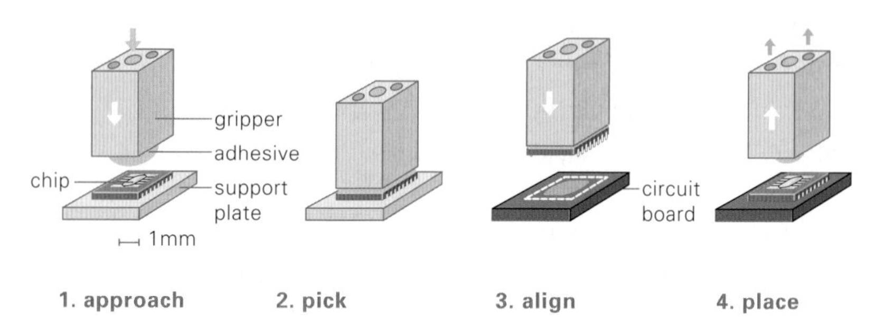

Figure 3.6 Adhesive gripping process (source: Bark)

Adhesive grippers have been developed for picking up micro components, e. g. by using the surface tension of a liquid drop to pick and place a chip. An accurate drop size is supplied onto the gripper`s underside and then slightly reduced during "gripping". The most difficult part of adhesive gripping is releasing the workpiece. In order to precisely maintain its position a workpiece must already be in a position where a force can be built up to counteract the adhesive force of the gripper. In figure 3.6 the gripper has already been provided with an adhesive. This counteracting force together with liquid reduction enable the workpiece to remain in the desired position. A gripper can also place workpieces with the help of an active release mechanism. Liquids are often volatile substances, e. g. alcohol, which do not leave any trace on the workpiece.

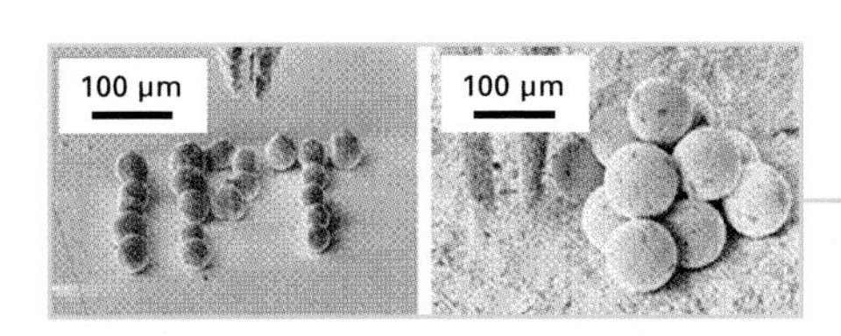

Place operation in the macro range (source: IPT)

Adhesion may occur due to unwanted adhesive forces between two different materials. In general these forces are neglectable and do not influence the gripping process in the macro range. However, adhesive forces may become critical when handling very small or delicate workpieces.

In the micro range strong forces due to electric charge occur on top of adhesive forces. The Scanning Electron Microscope (SEM) shows that glass balls with an approximate 100µm diameter are even able to move up along the gripper finger without additional manipulation. They are charged by electron bombardment in the SEM. This effect does not occur if gripper fingers and workpieces are grounded.

Cleanroom conditions prove the strong influence of humidity on adhesion. Minimum atmospheric humidity already reduces adhesive force. Surface quality also plays an important role for handling small component parts.

Since the adhesion makes the release more difficult, various techniques are employed to aid in the workpiece release. For example, wipe-off or vibration solutions are recommended by the Fraunhofer IPT in Aachen, Germany.

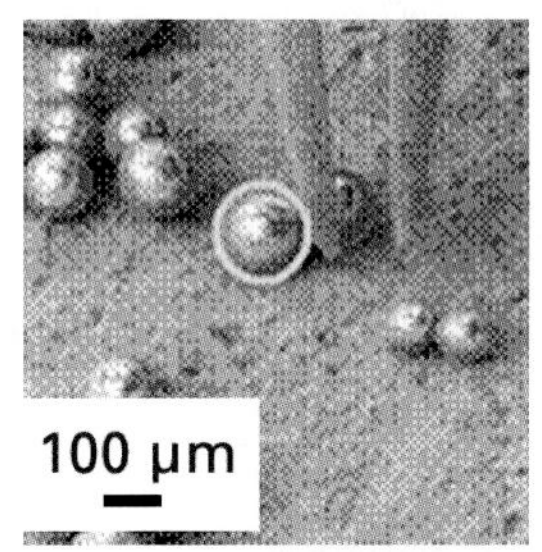

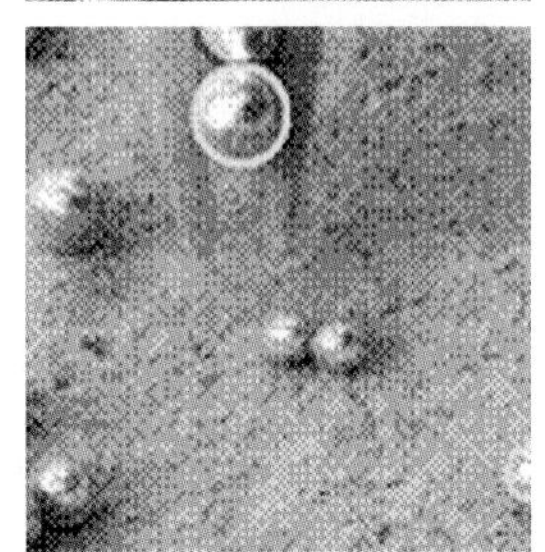

SEM image of glass spheres moved by electrostatic charge

The problematic point is that workpieces cannot be placed accurately because they remain stuck to the gripper because of adhesive forces. Therefore, gripper design needs careful attention to prevent this unwanted effect.

In the macro range such unwanted "adhesive effects" may arise from sticky workpieces. Special gripper fingers are required for such items in order to place them safely. Candied cherries, for example, will always stick on one of the gripper jaws and thus do not permit accurate release. It is not possible to foresee which gripper jaw a candied cherry will stick to.

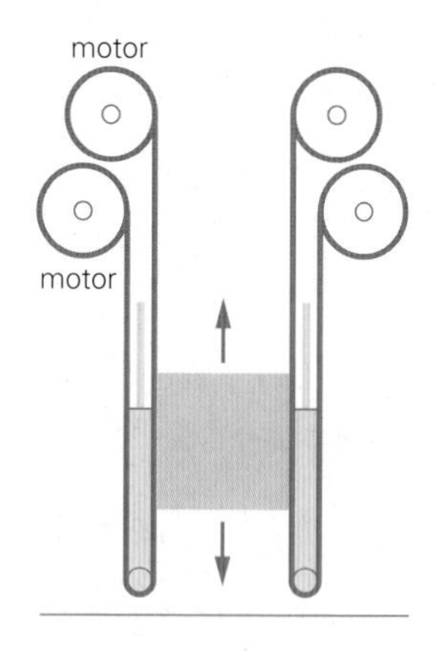

Figure 3.7 Flexible gripper finger – principle of function (source: SRI)

For this particular handling task special gripper fingers had to be fabricated. Even the operating elements of the gripper had to be driven by actuators to release the candied cherries. The cherry gripper developed by the SRI`s Mechatronics laboratory in the U.K. is able to pick, place, and pack sticky or delicate items. Its gripper jaws are unlike standard operating elements. A strip of moveable tape is wrapped around each finger of the gripper. The workpiece is released by retracting the fingers upwards and remains stationary while the tape progressively peels away. Being able to release the workpiece without opening the gripper allows packing in narrow spaces.

It is not only the food processing industry which has to deal with sticky workpiece surface. Grippers in the metal-working industry are regularly faced with coatings which may cause workpieces to stick. Handling workpieces the surface of which may be soiled, e. g. with resinified grease, is even more complicated.

sticky workpieces

Another example of workpieces making automated handling difficult is baggage. A gripper system has been developed especially for the purpose of stacking baggage into air-cargo containers.

The challenge of this task is to handle baggage of the most diverse sizes, shapes, and weights with one and the same gripper. Moreover, the gripping system needs to be designed for placing baggage into the container without restrictions to how and where it is placed.

These ambient conditions cannot be met by a standard jaw gripper nor by a kind of baggage shoveling system.

An innovative solution by KUKA InnoTec offers the possibility to attach the baggage to the gripper instead of having it "gripped". The baggage is strapped to the gripper with strings which are lashed in order to safely attach the workpiece to the gripper.

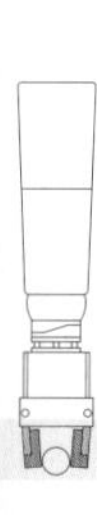

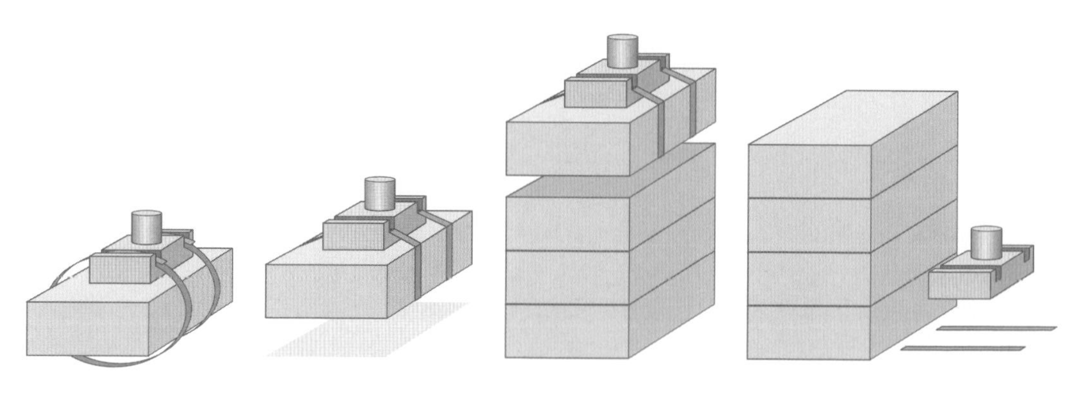

Figure 3.8 Baggage gripper – principle of function

This so-called static baggage gripper copes with any sizes, shapes, or weights, and is able to place the baggage in nearly any position. The principle of function is attaching the workpiece to the gripper by centering the gripper on the baggage and bracing both with two ties. By tightening the ties the baggage is properly fastened to the gripper. The patented gripping system is illustrated in figure 3.8.

The baggage is braced by a type of machine similar to those used for packaging parcels or bundles. The ties are guided by slides on two sides of the gripper to ensure lateral guiding and prevent the ties from slipping out. The ties are lead underneath the baggage to the other side and taken up by the bracing machine, which cuts the ties after lashing and welds their ends together. In order to prevent the ties from slipping within the guiding channels of the gripper they are clamped. Thus the baggage is safely attached to the gripper and ready to be placed into the air-cargo container.

Static baggage gripper with framed trunk "within" the gripper

After placing, the gripper cuts the ties with an inbuilt knife, rolls them up, and throws them out to dispose of them before the next pick- and place operation.

This is a novel type of gripper jaw design because bracing requires the ties to be thrown out after having placed the workpiece.
The drawback of these "throw-out jaws" is the necessity to dispose of waste material but using recycling materials helps to achieve a certain degree of materials flow.

Manually sorting recycling goods from packaging waste

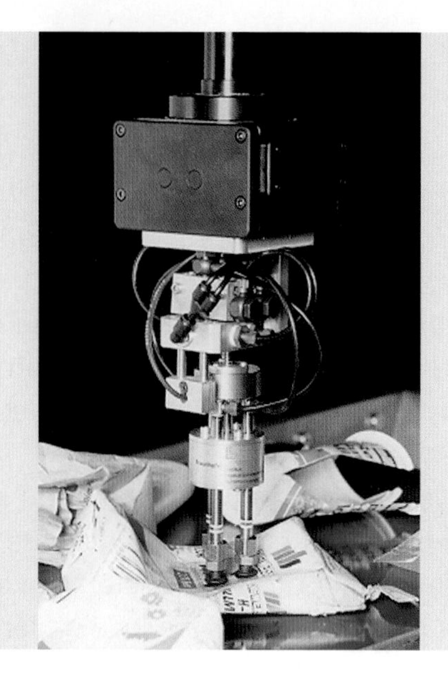

Grippers for foreign or hazardous materials can deform workpieces

The problem of waste material leads to a further option of gripper finger design. Handling tasks include sorting out defined parts from a flood of materials. The workpiece may be deformed or destroyed during handling as it will be crushed for recycling purposes in any case.

Collecting recycling materials always aims at generating pure quality materials flow. Materials are prepared by filters or magnetic separators, e.g. to leave just one kind of synthetic material of a defined quality within the materials flow. In order to recycle these synthetics they need to be as pure as possible without foreign or hazardous materials. In daily practice recycling categories are often faulty and materials still have to be sorted out by hand. Grippers which are able to cope with such tasks do not have to be adapted to any workpiece geometry. Usually it is sufficient that they pick the workpiece up regardless of whether it is being destroyed or damaged during handling.

3.3 Securing The Workpiece

For applying forces to the workpiece through contact surfaces the VDI Guideline 2860 describes the Secure function which includes the basic functions Hold and Release. Hold means temporarily securing the workpiece in a defined orientation and position, Release is the reverse of the Hold function.

If gripping the workpiece is effected by friction lock these basic functions become more complex. According to definition Clamp stands for a more complex Hold function and Unclamp stands for a more complex Release function.

Handling function	Symbol	Description of function
Hold		Hold means temporarily securing the workpiece in a defined orientation and position
Release		Release is the reverse function to the Hold function
Clamp		Clamp means temporarily securing the workpiece in a defined orientation and position under force-fit gripping
Unclamp		Unclamp is the reverse function of the Clamp function Note: Hold and Release are basic functions. Clamp and Unclamp correspond to Hold and Release on a more complex level

Table 3.15 Handling functions for securing the workpiece

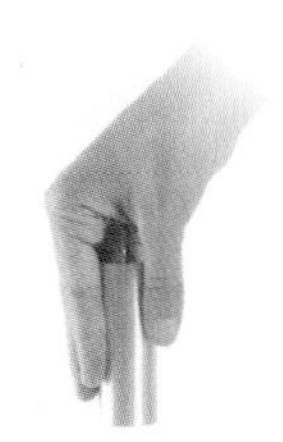

The human hand offers a whole range of options for gripping a workpiece. It can be gripped with either a strong or a soft grip. Variants of human gripping are illustrated here. Starting with the **cylinder grip**, which uses nearly the whole palm for applying force and thus offers maximum gripping safety, all the way through to indicating and operating functions which do not include actual gripping movements. With the help of these examples and your own gripping experience you can easily name options for force induction on a workpiece.

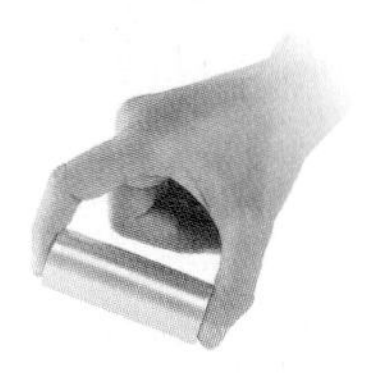

The **cylinder grip** for maximum force transmission is applied by the weight-lifter gripping the pole of the dumb-bell. The **three-point grip** is used by a person taking a bottle out of a beverage case. The **precision grip** permits very fine-tuned operations such as inserting a CD into a PC drive.

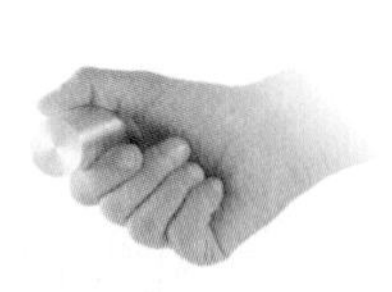

With the **lateral grip** a workpiece can be transferred to a new position or to another person. The **wedge grip** is a type of form-fit gripping which we all know from carrying a shopping bag.

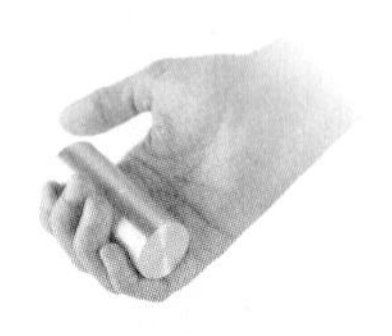

Indicating and operating functions of the hand are not rated as classic gripping actions but have become an essential part of the human environment, e. g. for operating keyboards.

Gripping options of the human hand (from top to bottom): cylinder grip, three-finger grip, precision grip, lateral grip, wedge grip, indicate/operate

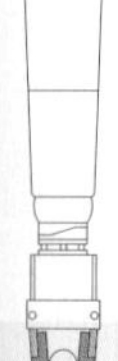

During a movement in the workspace both the direction of movements and the direction of forces acting on the workpiece change. Securing the workpiece requires provisions for changing gripping types during movements as shown in figure 3.9 to 3.11.

Pure form-fit gripping means that the active forces and mass moments of inertia are vertically directed to the contact surfaces between workpiece and gripper fingers during the entire process. Form-fit gripping is the best choice as gripping force can be kept low. Nevertheless, the direction of movement needs to be chosen carefully to make sure that the gripping type does not change during workpiece handling. Otherwise the defined gripping force may not be sufficient.

Pure force-fit gripping means that the gripping force of the gripper fingers is induced by friction force only. Thus gripping force is clearly dependent on the adhesive friction coefficient which may change under real working conditions, such as in dirty environments.

Force-fit gripping may also change to form-fit gripping because the gripper may change its direction of movement.

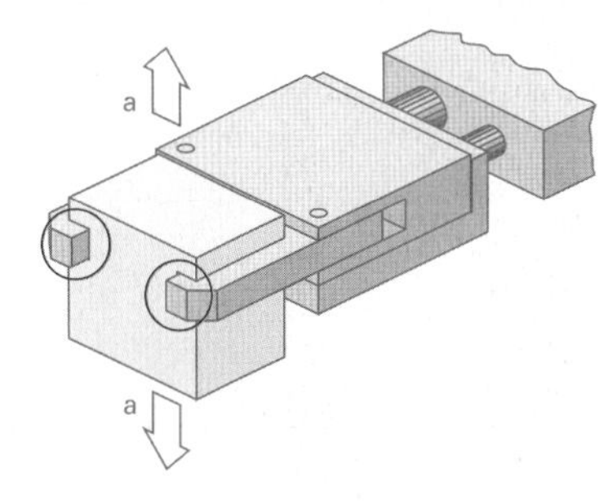

Figure 3.9 Form-fit gripping

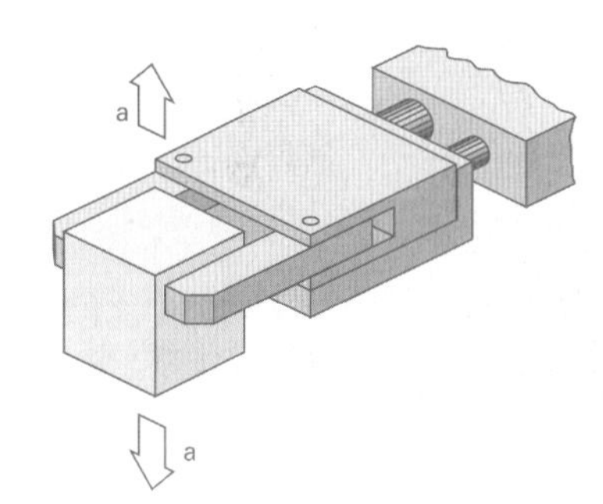

Figure 3.10 Force-fit gripping

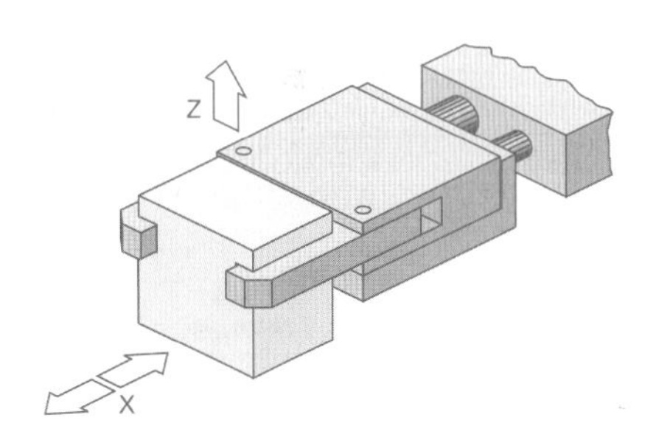

Figure 3.11 Form-fit gripping

A survey of these circumstances and corresponding examples calculation are given in table 3.16.

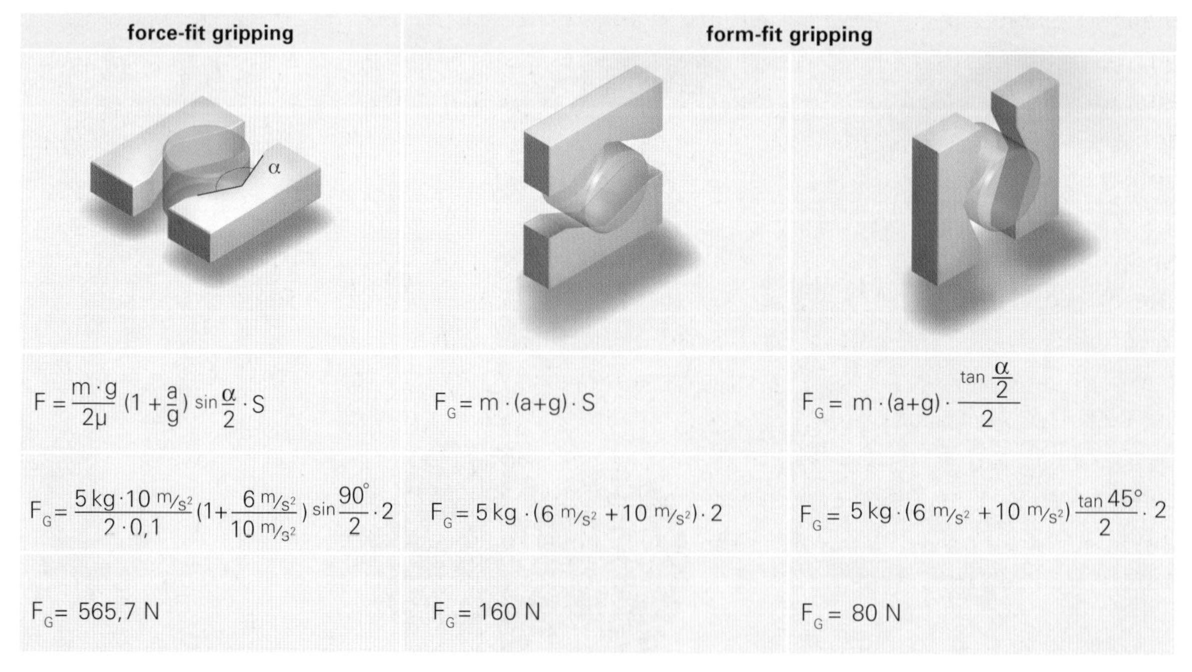

force-fit gripping	form-fit gripping	
$F = \frac{m \cdot g}{2\mu}\left(1 + \frac{a}{g}\right) \sin\frac{\alpha}{2} \cdot S$	$F_G = m \cdot (a+g) \cdot S$	$F_G = m \cdot (a+g) \cdot \frac{\tan\frac{\alpha}{2}}{2}$
$F_G = \frac{5\,kg \cdot 10\,m/s^2}{2 \cdot 0{,}1}\left(1 + \frac{6\,m/s^2}{10\,m/s^2}\right) \sin\frac{90^\circ}{2} \cdot 2$	$F_G = 5\,kg \cdot (6\,m/s^2 + 10\,m/s^2) \cdot 2$	$F_G = 5\,kg \cdot (6\,m/s^2 + 10\,m/s^2) \frac{\tan 45^\circ}{2} \cdot 2$
$F_G = 565{,}7$ N	$F_G = 160$ N	$F_G = 80$ N

Table 3.16 Force-fit/form-fit gripping

(S = safety factor)

As shown in table 3.16 a spherical workpiece is clamped between two prismatic jaws with the opening angle α. The gripping force is calculated with the appropriate formule. In the first case we need to consider the fact that weight acts against acceleration. The latter and the resulting adhesive friction must be completely assimilated by force-fit gripping, which requires higher gripping force than in the other examples. Form-fit gripping in the direction of acceleration and of gravity is provided in the second example. The gripper must counteract the force which stretches the gripper fingers caused by acceleration. The third example just requires half as much gripping force because the latter can be equally distributed to both gripper jaws. This type of handling is especially kind on the gripper and the workpiece as forces can be kept low.

Kinematics or drives for grippers

Gripper fingers need to be set in motion to build up gripping forces on the workpiece. Grippers can be categorized by their principles of drive. Our overview shows that mechanical grippers constitute the main representatives of gripping technology. Suction grippers, i. e. grippers with one or more switch-on/switch-off contact surfaces, are being widely used for industrial applications. Magnetic grippers, adhesive grippers, mold grippers, and needle grippers are still exceptions to the rule.

Mechanical gripper kinematics must transfer movement of drive into movement of gripper fingers, e. g. rotary drive movements have to be transferred into linear gripper finger movements. This type of kinematics includes all drive, transmission, and guiding elements which are necessary to realize the movement of the drive. The kinematic scheme shows most gripper drives in use, categorized by their input or output movement. The input movement is contrasted to the output movement for rotary and linear movements respectively.

mechanical grippers		**suction grippers**	**magnetic grippers**	**adhesive grippers**	**mold grippers**	**nail grippers**
scissors gripper	fork gripper		electromagnet	adhesive foil		
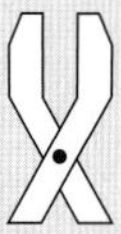		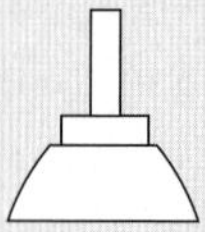	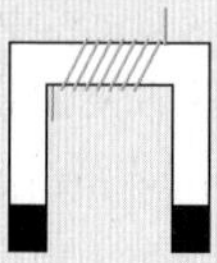		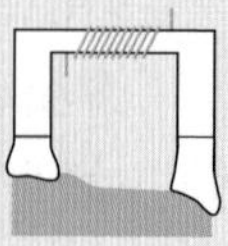	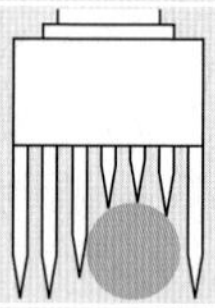
parallel jaw gripper	three-point gripper		permanent magnet			
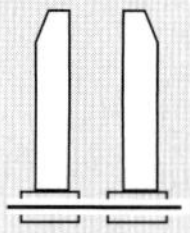	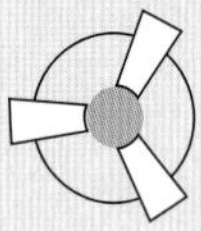		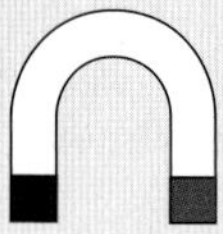			

Table 3.17 Gripper types categorized by their principle of drive

		input movement										
		linear						**rotary**				
		shear grinding drive	tenon drive	fork lever drive	clip lever drive	wedge drive with rocker switch	elbow lever drive	curve tongue drive	anchor drive	thread spindle drive	excenter drive
output movement	**linear**	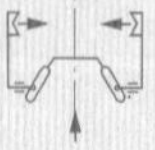	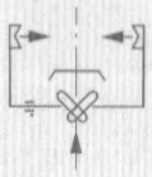			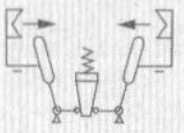			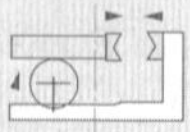	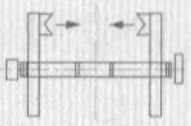	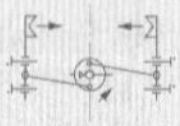
		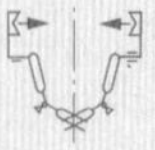									
	rotary			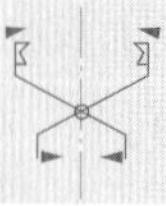	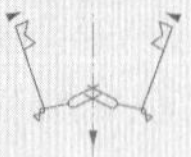	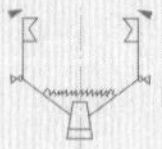	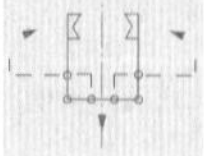				
							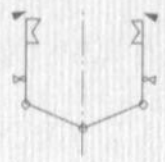				

Table 3.18 Comparison of input and output movement of gripper fingers related to the respective type of kinematics (source: Dreher)

Inputs and outputs are related to the respective types of kinematics, for some types two different solutions are possible.

The gear ratio characterizes gripper kinematics, as it describes the ratio of velocity of drive to the velocity of gripper fingers. The drawback of lever drives, especially the elbow lever drive, is that the area where high clamping forces are reached is very small. Therefore, a gripper with elbow lever drive must be specially designed for a particular workpiece size. With tenon and shear grinding drives either a constant or a distance-related distribution of gripping force over clamping distance can be achieved in relation to the type of construction.

Kinematics with rotary jaw movement can be produced for high process reliability at low cost due to their simple construction. Moreover, this principle of kinematics achieves large travel (swivel movements) of gripper fingers even for small gripper sizes. Workpieces with contact surfaces parallel to each other require gripper active surfaces to be flexible to avoid insufficient point contact with the workpiece.

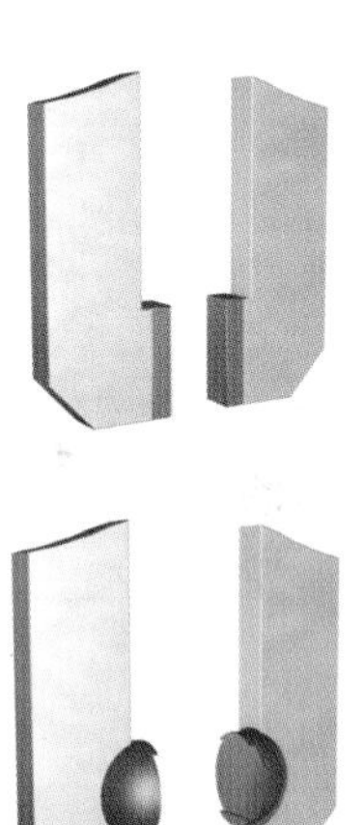

A simple method to ensure plane surface contact

Different contact surfaces may lead the gripper to loose the workpiece in case the force which can be transmitted becomes too low. Surface contact of the gripper jaws may change to insufficient point contact for varying workpiece dimensions. Selecting the right gripper includes making sure that the movement types of the operating elements are suitable for the respective workpieces. If certain ambient conditions demand a type of movement which may result in contact surface tolerances during gripping, swivel elements in the fingers or compensation units can set them off.

Positioning tolerance makes it even more difficult to position cylindrical workpieces of varying diameters within prismatic jaws. This tolerance caused by the circular movement of the gripper jaws does not occur when gripper fingers are shifted in parallel.
The figures show different movements of gripper fingers according to the respective kinematics employed.

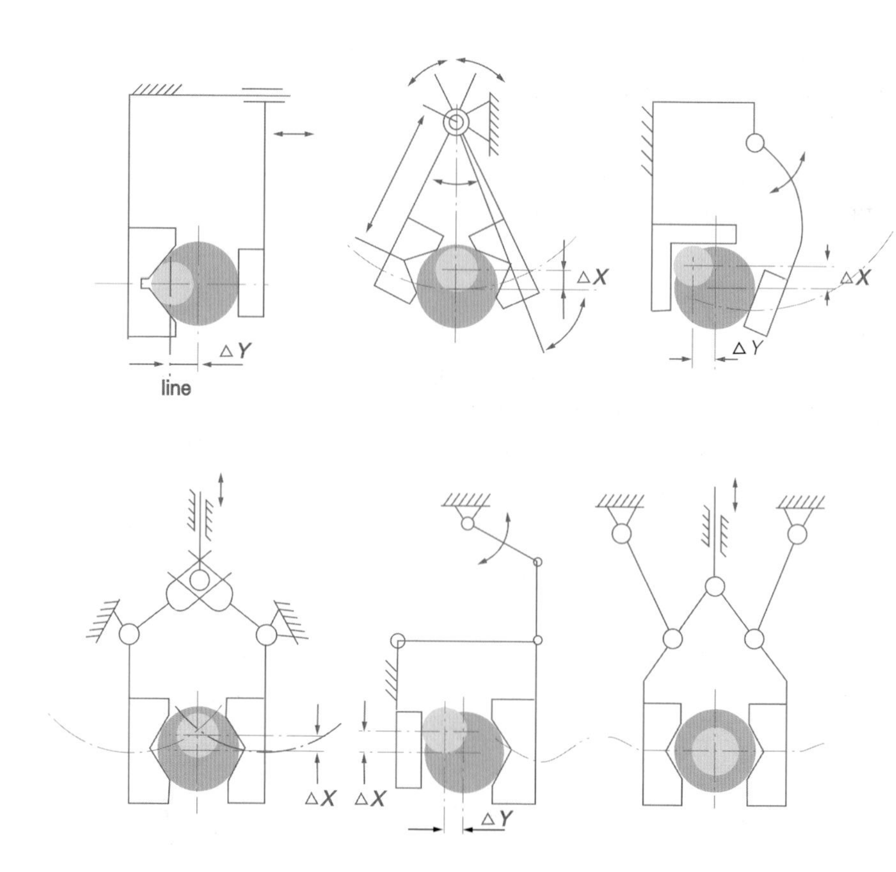

Figure 3.12 Movements of gripper fingers according to type of kinematics

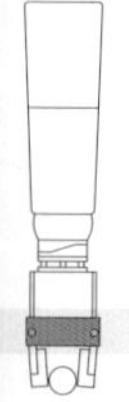

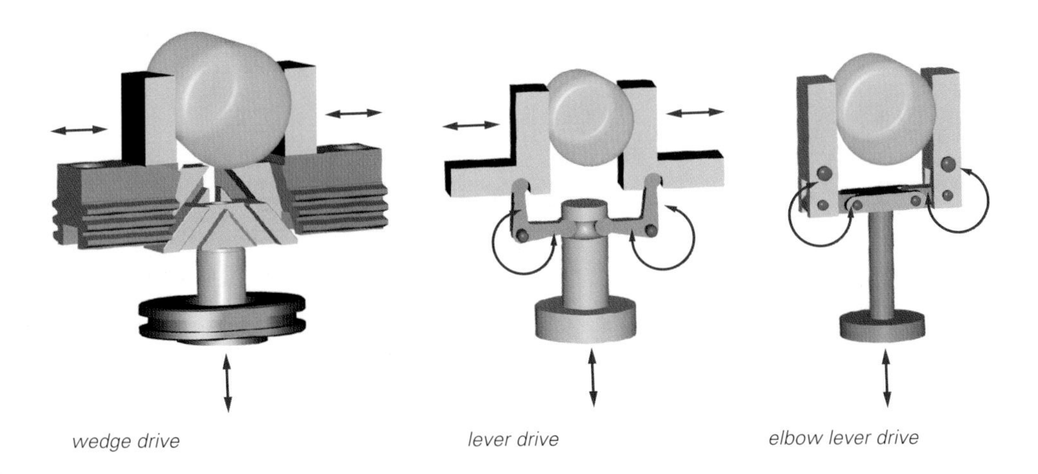

Gripper fingers with parallel guidance are better suited for work-pieces with parallel contact surfaces because line contact with the gripper jaws can be avoided in case of tolerances. Preferred kinematics with their vital characteristics are illustrated and explained in the following.

The wedge drive is noted for a constant distribution of force over the entire stroke length. It is especially suitable for gripping work-pieces of differing dimensions or great tolerances with the same kinematics. The constant force ensures that the workpieces are picked up under the same conditions. This gripper type also stands out for high precision gripping. Guidance and friction at the wedge cause forces which have to be counteracted during movement.

The lever drive is distinguished by low friction forces in the lever joints. However, it does not allow precision gripping at the level of the wedge drive and force over the entire stroke is unstable.

The same applies to the elbow lever drive with the difference that very high forces can be achieved in its final position. However, it is not flexible in the face of workpiece tolerances so that greatly differing workpiece dimensions require specially adapted kinematics.

The workpiece is gripped by building up a force through kinetic movement which translates the driving power of an actuator into a rotary or translatory movement. Actuators are employed as subsystems for all grippers which use kinematics for moving the gripper fingers. The drive is responsible for transforming the energy supplied into a rotary or translatory movement. As a rule, the drive is directly connected to kinematics.

This stationary connection must be given up when the workpiece needs to be gripped by several fingers, as it is the case with the Barret Hand. For delicate finger structures it is essential that stress on transmission or drive at gripper closing is kept low. Otherwise the service life of such lightweight constructions would be significantly shortened. A special driving gear is necessary which allows the gripper fingers to be aligned with the workpiece without putting too much strain on the driving components. A so-called decoupling drive is used which makes it possible to use one driving motor per finger. When the gripper clamps a workpiece both finger joints are closed. When the first joint meets resistance the drive of this joint is decoupled by a plate spring. The second joint continues moving until it is aligned with the workpiece. The motor is subsequently switched off by gripping force control.

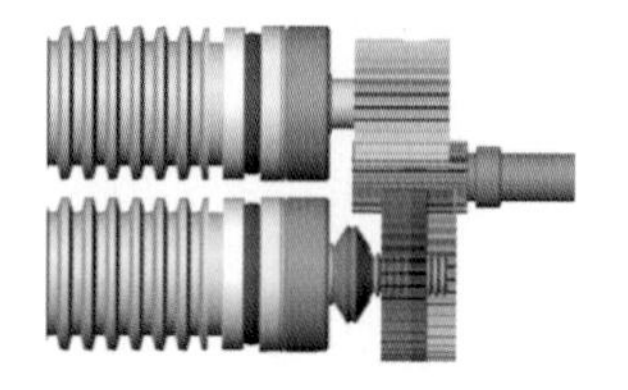

position "link 1 adjusted" (red), position "workpiece gripped" (blue)

Passively or actively adjusting the operating elements of the gripper is a good option for gripping the most diverse workpiece geometries. The number of fingers also determines the degree of flexibility for different gripper concepts of artificial hands as illustrated for different artificial hands in Chapter 2.

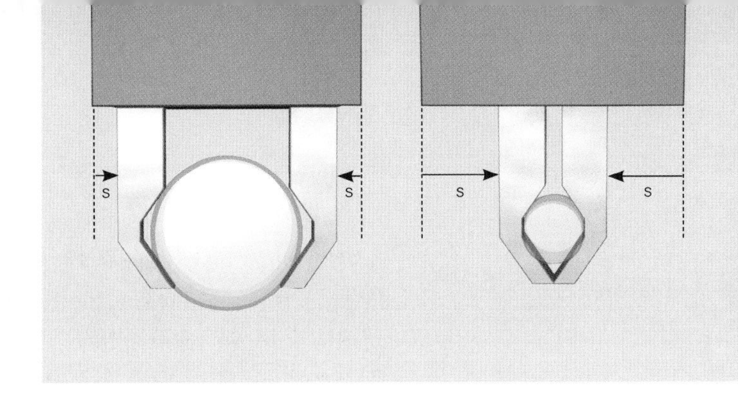

Picking up different material diameters with a prism illustrates the role of the finger stroke (s) for gripper flexibility

Grippers can be categorized by the number of their operating elements. The more operating elements a gripper has the more options to apply forces on a workpiece it has. However, realizing these options starts with designing the operating elements, i. e. the gripper fingers.

Gripping various diameters of cylindrical workpieces can be realized by a uniform prism design if the finger travel is long enough. The stroke or the distance a gripper finger is able to travel, can be decisive for gripper flexibility. The latter can be determined by easy CAD analysis of overlapping workpiece geometries in order to handle as many different workpieces as possible with one set of gripper jaws.

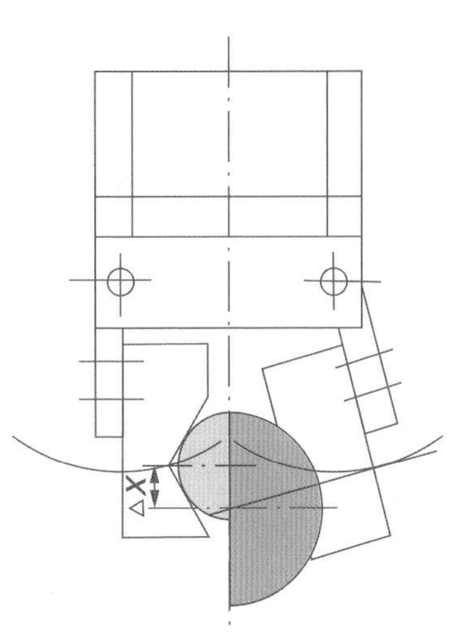

As shown in the example, the crucial point of such an application is the differing center point position of the tubes when gripping them with the same prism. Pick- and place operations of tubes must be carefully planned to avoid faulty positioning.

If very high flexibility is required, systems for changing gripper fingers are a solution. Within such a system the operating elements of the gripper are replaced to be able to meet a new handling task. The gripper fingers can either be changed manually or automatically. The latter is required for very quick changing times. An automated quick-change prevents the gripper jaws from getting mixed up, which always remains a risk of manual change. Automated change systems can be categorized according to whether fingers can be changed "on the fly" during handling operation or whether they are directed to a special position where they are changed.

Changing gripper fingers "on the fly" means that the kinetic device is able to continue its movement without being interrupted.

option in terms of economic and technical requirements. If gripper fingers are changed too often, production will become too expensive. Quick-change jaws, however, can even bring a manual change of gripper fingers down to a few seconds. As mentioned earlier, manual change always includes the risk of gripper jaws getting mixed up by human operators. This may result in faulty production or even in a collision between workpieces and the wrong gripper fingers. Under these aspects an automated change of gripper fingers will add to safety provisions in production.

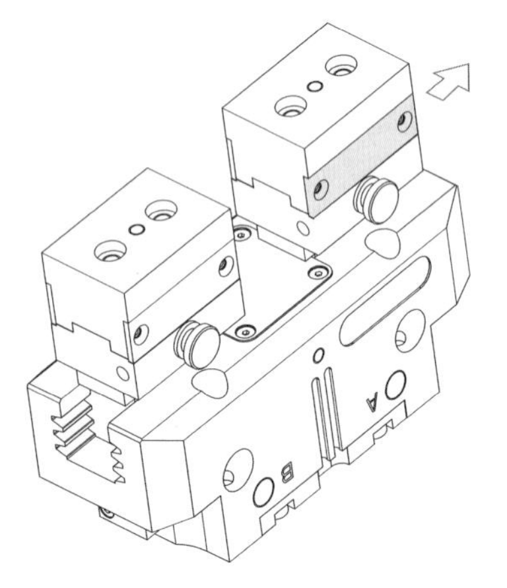

change unit for gripper fingers

Another method of increasing gripper flexibility is the use of so-called swivel or rotary units. They combine several gripper finger types on one swivel or rotary plate.

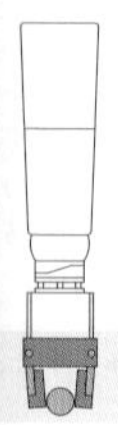

dual gripper (above) and revolver gripper (left)

With these units different workpieces can be handled without changing the gripper jaws of a handling system. The rotary unit takes the respective gripper with suitable gripper fingers into the right position for picking up the workpiece.

The swivel or rotary units can be designed as dual grippers or revolver grippers. The difference between the two is the number of grippers one unit can hold. Our focus is on dual gripper constructions most frequently used for handling workpieces in order to synchronize handling time and processing times.

A dual gripper can hold two grippers which are designed to operate independently of each other.

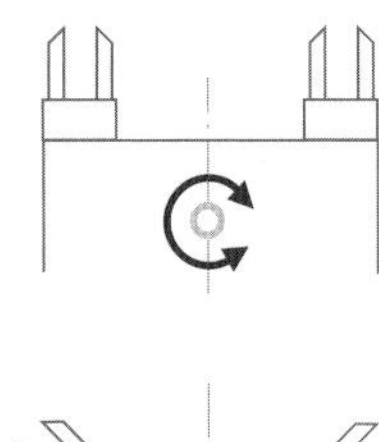

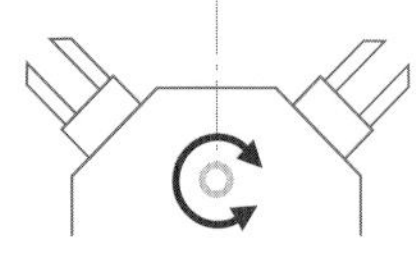

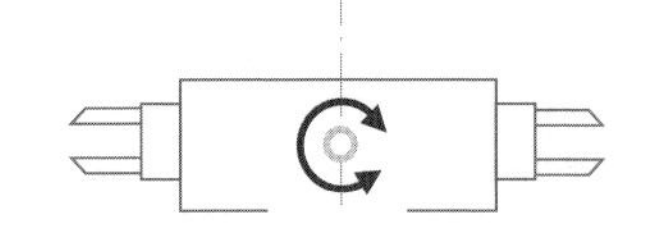

parallel, coniform and radial structure of revolver grippers

A revolver gripper consists of more than two grippers which are able to work independently and is predominantly used for handling several workpiece types. One workpiece type is distinguished from another according to which gripper is able to cope with it. The structure of the operating elements on a dual gripper or a revolver gripper may be parallel, coniform or radial.

Small and medium-sized product lines demand gripping technology to be even more flexible as the aim is always to cover the broadest range of workpieces possible. Gripper fingers with a long stroke meet this demand.

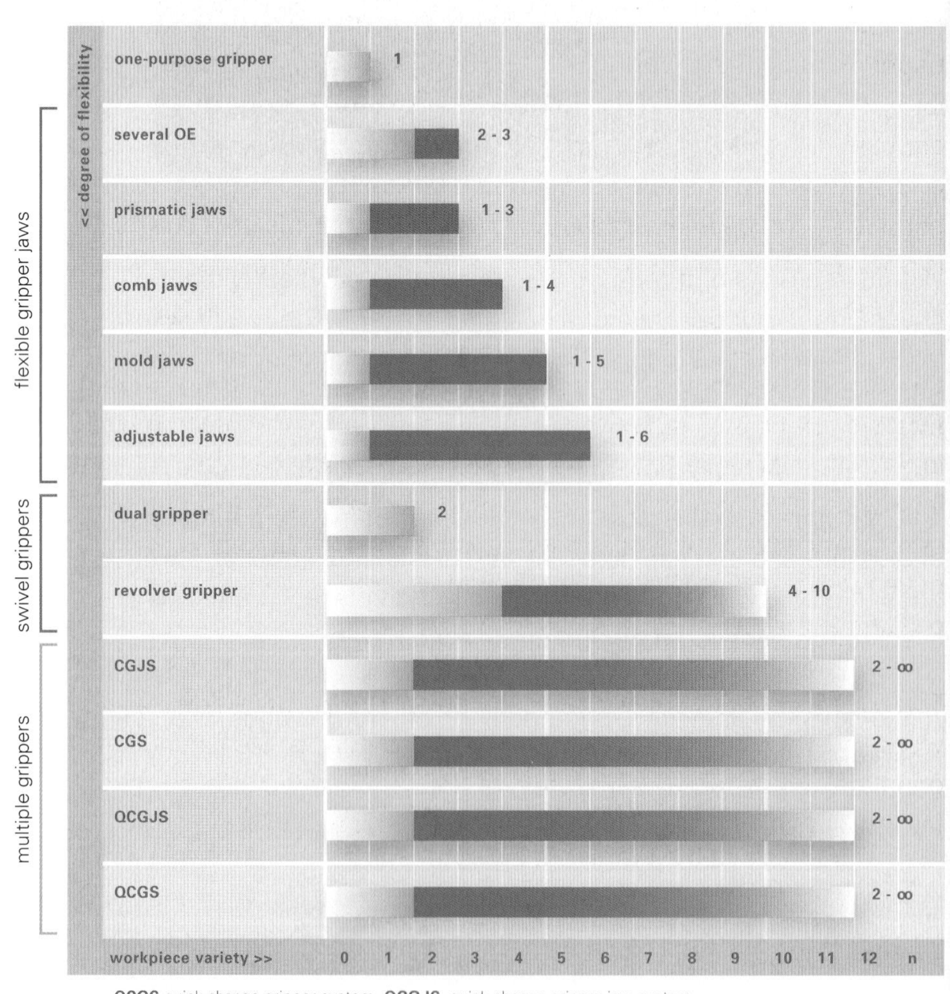

QCGS quick-change gripper system **QCGJS** quick-change gripper jaw system
CGS change gripper system **CGJS** change gripper jaw system **OE** operating elements

Figure 3.13 Flexible gripper systems coping with workpiece variety

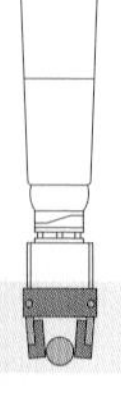

Nevertheless, workpiece variety is on its way into large series production as well. Mass production in its true sense is only relevant for single workpieces, components, or whole systems. Consumer goods, e. g. automobiles, are increasingly produced according to individual order. Lower lot numbers again require more handling flexibility. Robots in combination with flexible gripping technology and sensors are the latest state of engineering. Expenses for the respective high-tech components rise according to the performance required.

Figure 3.13 explains various options for gripper systems coping with workpiece variety. The single purpose gripper has gripper jaws tailored to the workpiece and can be equipped with flexible gripper fingers for additional flexibility. A gripper system`s degree of flexibility is expressed by the quantity of different workpieces this gripper system can cope with. The first step is to use several contact surfaces and prismatic jaws compared to the single purpose gripper. In combination with comb, adaptable and adjustable jaws it is possible to grip most different workpiece geometries.

For single purpose grippers mounted to swivel or rotary units it is the housing which determines the number of grippers that can be fitted to one unit. Revolver grippers definitely offer the best options in terms of flexibility. Multiple grippers theoretically handle an infinite number of workpiece geometries by either just changing gripper fingers or the whole gripper unit including kinematics and drive (see details in Chapter 4). The latter opens up additional options with regard to finger stroke or type of drive.

The distance which the gripper fingers cover to apply force on a workpiece, the so-called action radius of the gripper, also influences gripper flexibility. The workpiece size and gripping type define how far the operating elements of the gripper need to be opened.

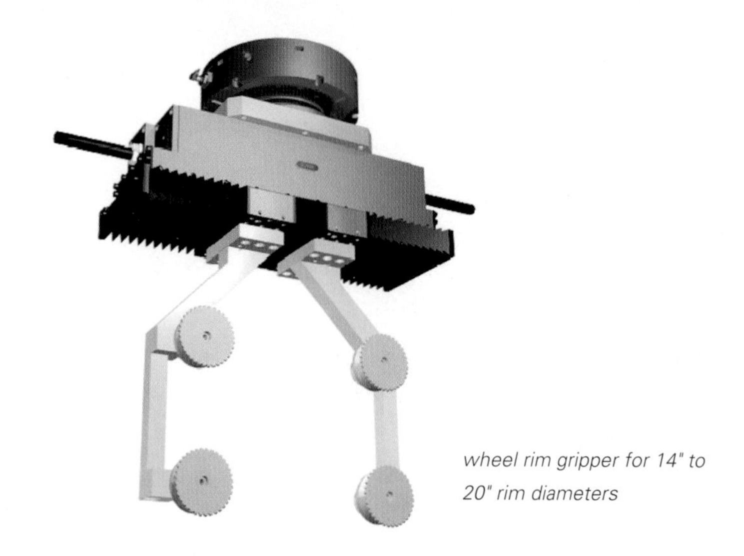

wheel rim gripper for 14" to 20" rim diameters

The larger a gripper`s action radius is, the more workpiece variety it can handle. It is not the maximum gripping radius which is important but gripper finger travel.

Handling wheel rims is a good example because they are produced in differing diameters. The action radius necessary for this gripper covers 14" to 20" wheel rim diameters. Nevertheless, enlarging gripper finger travel to cope with a larger range of workpieces may backfire in terms of gripper housing, gripper weight, and closing time. Set cycle times for time-critical applications may not be met in the end.

This dilemma can be solved by a gripper concept with gripper fingers being pre-positioned before the actual gripping process so that the distance the gripper needs to cover for supplying the gripping force is much shorter. Electrically driven gripper fingers are ideal because finger positioning can be programmed.

Various gripper drive types can be categorized according to their respective principle of function. In table 3.19 current gripper drive types are compared. Electrically and pneumatically driven grippers cover a broad range of handling tasks while hydraulic drives are predominantly used for grippers handling high payloads. The piezo-electric drive is rarely used and generally reserved for gripping technology in the micro range due to its particular gripping force and gripper finger stroke. The best gripper principle of function always needs to be selected in relation to the specific handling task.

The pneumatic drive stands out for its simplicity and long service life, good-quality air pressure for it is usually available in production workshop environments. Pneumatics enable compact housing of the drive element. This type of drive is protected against overload by compressible air pressure. Pneumatically driven grippers are able to cope with extreme conditions, e. g. coolants or dust from casting or grinding processes. Moreover, these drives reliably operate in powerful electric or magnetic fields. Another benefit is fast opening and closing times. In comparison to other types of drive pneumatic drives are a very low in prime costs and save energy costs. Additionally, these drives have the feature of being explosion-proof.

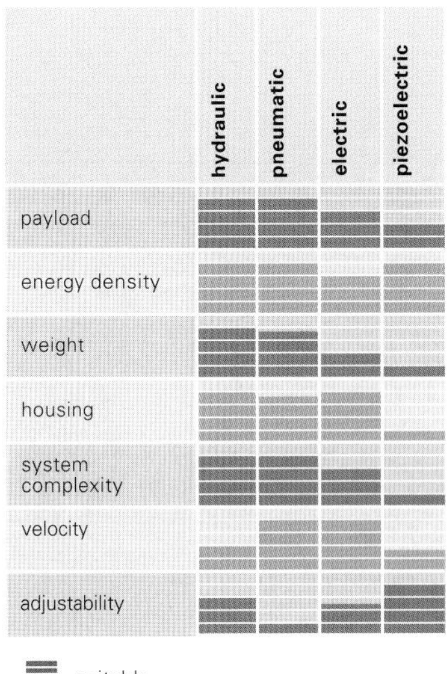

Table 3.19 Principles of gripper drives and their performance features (source: Fraunhofer IPA)

Adjustability of pneumatics is very limited compared to other types of drives. Waste air which is drawn off directly from the gripper is to be treated separately for special applications in cleanroom or strict hygiene environments. Pneumatic drives frequently require final position stabilizers to avoid damage in case the gripper moves too hard into its final position. The noise level of pneumatic drives is higher than that of other types of drives.

The hydraulic drive can transmit great forces despite small housing. Moreover, it permits an infinitely variable regulation of constant velocity of travel and gripping force can be upheld over the entire gripping path as well. Maximum force is achieved even at small distances because mass moment of inertia of the elements moved and compressibility of the oil are low.

One of the drawbacks of hydraulic drives is a cost-intensive servicing routine because leakage of the gripper or its supplies may lead to serious damage. Compared to other types of drive the energy supply is more complicated as hydraulic systems are rarely part of in-house technology for production. In most cases they would have to be purchased and installed separately. Recycling hydraulic oil for re-use within the circuit requires additional expenditure. Supplying energy to hydraulic grippers within robot systems equipped with manual orientation axes is far more difficult to realize than with any other types of energy. A delivery valve is necessary to limit gripping force.

Electric drives permit excellent control of generating force and movements, other advantages are low prime and operating costs. Compact construction of electromotors and improvements of efficiency have drawn more and more attention to electric drives for gripper technology over the past few years. Modern grippers with integrated sensor technology in combination with electric drives make direct gripping force control possible.

The piezoelectric drive is especially useful for small fast movements. This drive technology is characterized by high energy density and offers an excellent possibility to produce compact micro gripper drives. In terms of control piezoelectric drives are superior to pneumatic drives. Due to their low forces and small distances piezoelectric drives are limited to the micro range and do not cope well with workpiece variety.

Piezo gripper

Each principle of drive requires a transformation of the respective type of energy into movement by a so-called actuator. Actuators are used as gripper drive components. Gripper kinematics are driven by either translatory or rotary movements. Components of pneumatic drive technology are pneumatic cylinders, swivel cylinders, or air-pressure motors. Hydraulic cylinders, swivel cylinders, or hydromotors can be considered as drive components of hydraulic actuators as well. Drives based on the electric principle of function include electromagnets, piezo drives, linear motors, as well as rotary actuators such as stepping motors, direct-current (DC) and alternating-current (AC) motors.

	pneumatic	**hydraulic**	**electric**
translatory drive movement with limited travel	pneumatic cylinder	hydraulic czylinder	electromotor
translatory drive movement with unlimited travel			linear motor
rotary drive movement with limited rotary angle	swivel/rotary cylinder	swivel/rotary cylinder	
rotary drive movement with unlimited rotary angle	air-pressure motor	hydromotor	stepping motor DC motor AC motor

Table 3.20 Various gripper drives for different types of energy sypply

Selecting a gripper drive in relation to kinematics determines how the operating elements move in terms of gripping radius and velocity. This also specifies the type of gripping force which can be applied to the workpiece, and together with the type of gripper fingers it finally determines the principle of gripping, e. g. form-fit or force-fit gripping.

gripper-related characteristics

contact surfaces | gripping force | gripping time | gripping area

Gripper-related characteristics for pneumatically driven grippers, which are widely employed in industrial applications, are illustrated in the following.

Contact surfaces

The role of contact surfaces has already been explained in detail. Number and design of contact surfaces affect calculations of gripping force in terms of how this force is to be decomposed. If numerals are used within gripper names, such as 3-finger concentric gripper or 2-finger parallel gripper, they refer to the number of contact surfaces.

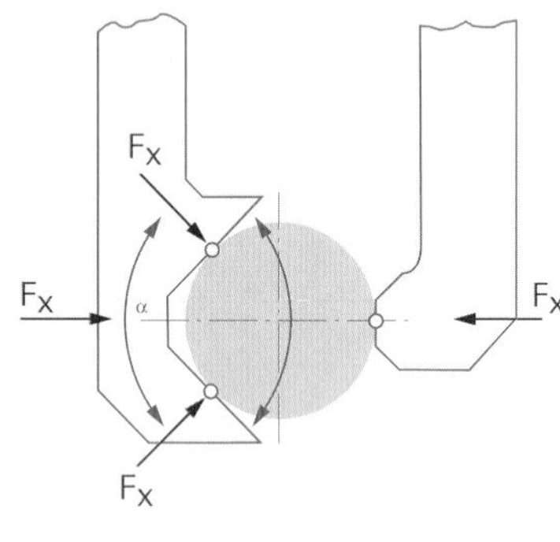

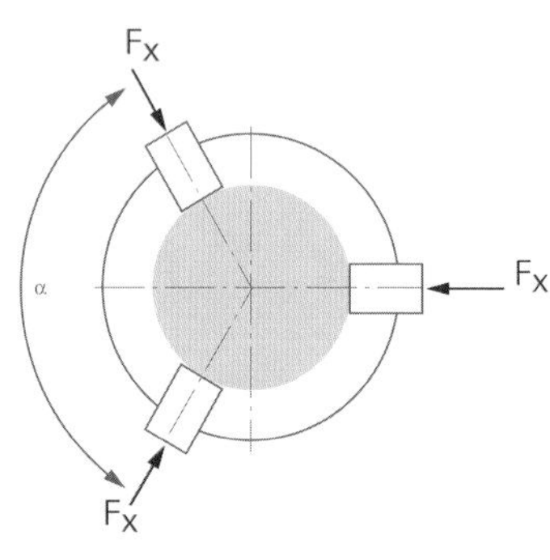

Figure 3.14 Gripping force is to be decomposed according to number of contact surfaces and number of fingers

Gripping force

The determining characteristic for many applications is the gripping force or the weight of the workpiece which the gripper is able to handle. As mentioned earlier, the required gripping force is first of all a question of which forces can be applied to which contact surfaces of the workpiece. Once the latter is established the required gripping force can be calculated with the formule as described. This characteristic defines a gripper`s force which the operating elements or gripper fingers apply to a workpiece.

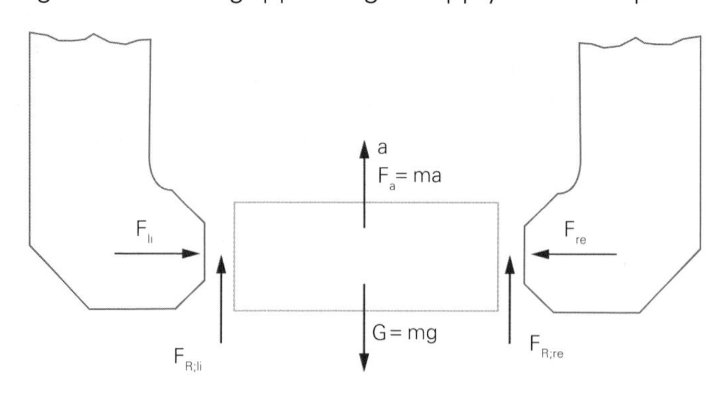

Figure 3.15 Decomposing forces on a workpiece for force-fit gripping

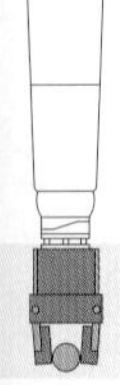

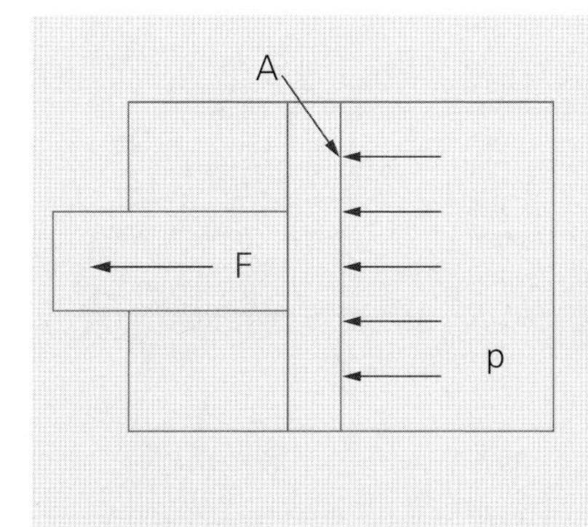

pressure = $\frac{\textbf{force}}{\textbf{surface}}$

$$\mathbf{p = \frac{F}{A}}$$

Example

$F = 2MN$; piston $\cdot \varnothing d = 400\ mm$; $p = ?$

$$p = \frac{F}{A} = \frac{2000000\ N}{\frac{\pi \cdot (40cm)^2}{4}} = 1591\ \frac{N}{cm^2} = 159{,}1\ bar$$

1 Pa $= 1 \frac{N}{m^2} = 10^{-5}\ bar$

1 bar $= 10 \frac{N}{cm^2}$

$= 100000\ Pa$

1 mbar $= 100\ Pa$

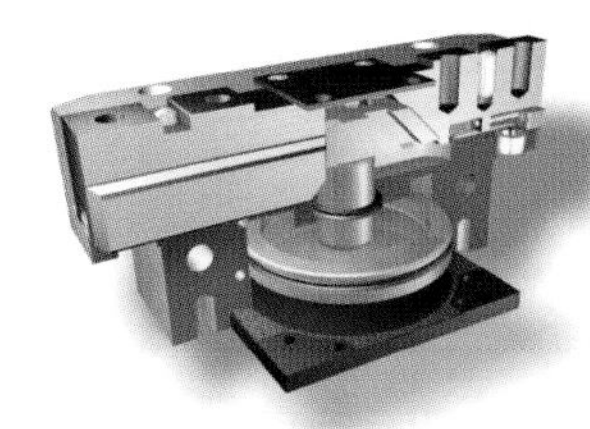

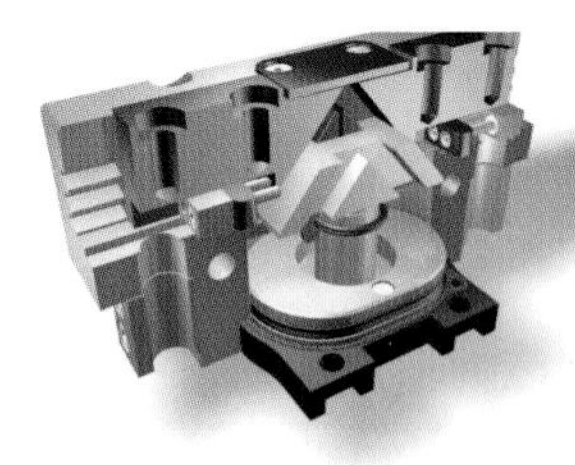

circular and elliptic piston surface

Pneumatically driven grippers normally use a piston to convert the energy saved in compressed air into a translatory movement.
The piston force is calculated as described. In modern pneumatically driven gripper systems even elliptic pistons are employed. This type of construction is ideal for exploiting the plane area determined by kinematics.

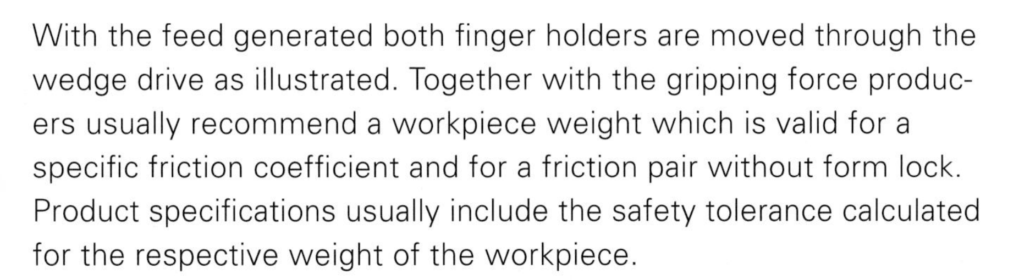

With the feed generated both finger holders are moved through the wedge drive as illustrated. Together with the gripping force producers usually recommend a workpiece weight which is valid for a specific friction coefficient and for a friction pair without form lock. Product specifications usually include the safety tolerance calculated for the respective weight of the workpiece.

Practical experience shows that it is important to know how the force is distributed over the length of the finger stroke.
In accordance with the kinematics used gripping force differs over the entire stroke. The gripping force diagrams in table 3.16 show that only the parallel jaw gripper with one wedge principle of function, for example, will achieve a constant distribution of force over the entire stroke.

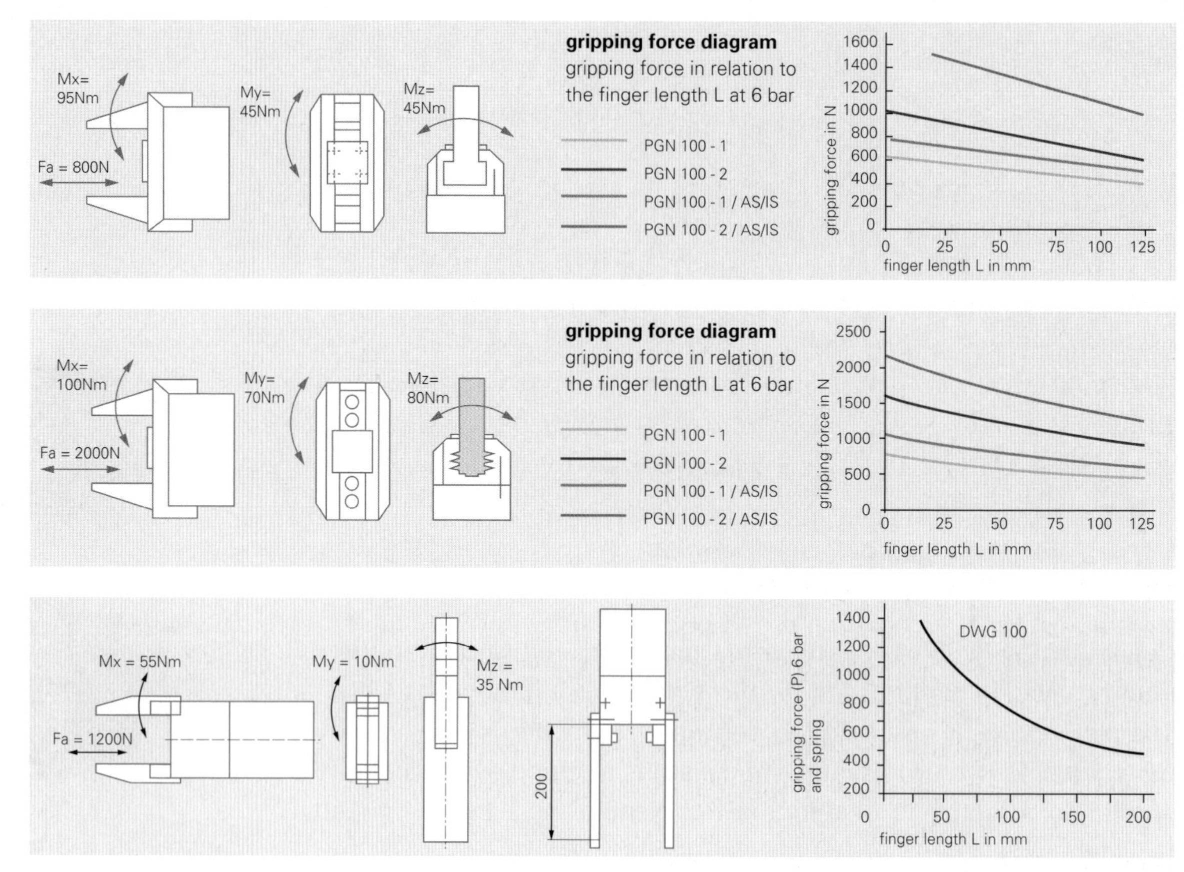

Figure 3.16 Different force distribution for various gripper types – maximum admissible forces and moments at the gripper fingers in addition to the gripping force.

gripper with serrated guides for increased moment capacity

The length of the gripper fingers influences the forces and moments occurring at the gripper kinematics. Therefore, gripping force is frequently specified in relation to the finger length in such a diagram to exclude overload or premature wear.

The characteristic curve for each gripper type shown in the gripping force diagrams falls with increasing finger length. Most evident is the difference between swivel grippers and grippers based on the wedge principle of drive. The gently declining curve of the PGN gripper and the nearly identical PGN plus 100 reflects high load capacity and robust guides for long finger capability.

The curve of angular grippers must obviously drop as in the example of the DWG 100 by SCHUNK, falling from a gripping force of 1,400N at 50mm finger length to a gripping force of 500N at 200mm finger length. This drop in gripping force, however, is not only a matter of straining guides and bearings of the gripper kinematics. The moment of an angular gripper, which is induced through the extended lever arm of a finger into the kinematics, counteracts the force of drive so that the piston must counteract the latter.

Opening and closing time of mechanical grippers

In most applications cycle time or process time for performing a handling task are essential for the efficiency of an automated solution. Part of the entire process time is taken up by opening or closing the gripper. Opening and closing times depend on the length of stroke, on the type of drive, and on gripper kinematics.

A gripper with gripping force maintenance (GFM) will have different opening and closing times as the spring force at opening must be overcome. When closing the gripper the spring will function as a support. As compared to other kinematics in table 3.21 the rack and pinion principle does have the shortest opening and closing times in relation to the stroke.

type of gripper	kinematics	drive	stroke	opening	closing
2-finger parallel	wedge principle without GFM	pneumatisch	4 mm	0.04 s	0.4 s
2-finger parallel	wedge principle with GFM	pneumatisch	4 mm	0.05 s	0.03 s
3-finger concentric	wedge principle	pneumatisch	4 mm	0.03 s	0.03 s
2-finger parallel	lever principle	pneumatisch	4.5 mm	0.05 s	0.05 s
2-finger parallel	rack and pinion	pneumatisch	15 mm	0.045 s	0.06 s

Table 3.21 Opening and closing times of various gripper constructions (GFM= gripping force maintenance)

Changing the pressure will only influence gripping force while opening and closing times remain the same for grippers without gripping force maintenance.

Gripping path

A successful pick operation requires the gripper to open in accordance with the form of the gripper jaws and the direction of the gripper approaching the workpiece. Therefore, the finger stroke necessary for the pick operation is called required jaw stroke c.

As illustrated in figures 3.17 and 3.18, a gripper needs to be opened further for a radial grip, i. e. when the gripper approaches the workpiece from the sides, than for an axial grip, i. e. when the gripper approaches the workpiece from above. In order to avoid collision, a gripper with a longer stroke of the jaws needs to be selected for the radial grip rather than for the axial grip.

The so-called clamping reserve b and the opening reserve a are distances to ensure a degree of safety with regard to workpiece dimension. If some workpieces should come out slightly smaller in diameter due to manufacturing tolerance the clamping reserve compensates for it. Thus the gripper is able to safely grip workpieces in case of smaller workpiece dimension. The opening reserve permits further opening of the gripper than necessary for standard workpiece dimensions in order to avoid collisions caused by oversized workpieces.

The stroke stated in product specifications for each gripper type may range from 4mm to 200mm for pneumatic grippers. It is important that the stroke is specified per gripper jaw. Frequently, grippers are specially classified as short-stroke and long-stroke grippers. As the name suggests short-stroke grippers are used for short opening and closing times or in case workpiece accessibility does not permit longer strokes.

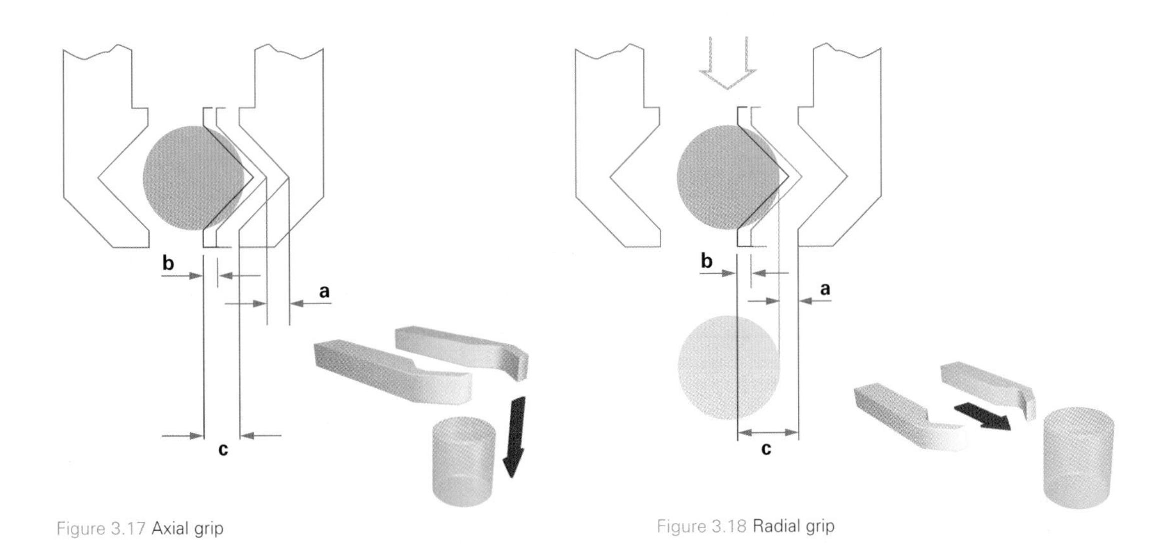

Figure 3.17 Axial grip

Figure 3.18 Radial grip

Their housing determines the application options of mechanical grippers because interfering edges must always be taken into account. Collisions with the gripper in open position occur every time the stroke has not been considered for or adapted to the size of the housing. Possible pick situations of different workpieces must be taken into consideration to avoid collisions. Long-stroke grippers cover a broad range of workpiece dimensions and can be used more flexibly for different workpiece sizes.

The decision for a particular gripper not only depends on workpiece- and gripper-related characteristics but also to a great extent on the ambient conditions of the pick operation.

3.4 Gripping Situations

As they strongly affect gripper design various gripping situations are described in the following. The focus is set on how workpieces are presented to the gripper for pick operations.

- picking up workpieces without order status ("grip at random")
- picking up workpieces with unsorted order status from a plane surface, e. g. a conveyor
- picking up workpieces with sorted order status, e. g. from a workpiece support

The above tasks may again vary according to whether workpieces have to be picked at rest or in motion.

For place operations the respective scenarios apply with one exception: Workpieces are rarely placed into an unsorted order status as they are usually desired in sorted order status.

Scenario 1: Workpieces Without Order Status

Picking up workpieces which are presented to the gripper without any order status is referred to as "grip at random". This expression already suggests that it is hardly possible to calculate all eventual collisions with the gripper jaws in advance. According to position and orientation of the workpieces lying in a box at random, the gripper fingers are faced with most different interfering edges of the workpieces. Therefore, this gripping situation requires sensors and subsequent safe actuation of the handling device. There are exceptions to the rule, e. g. if workpieces are made of elastic material and thus can be simply pushed aside by the operating elements of the gripper.

In an entirely unsorted situation hardly any automated system can cope. The "grip at random" has been repeatedly promoted and demonstrated at trade fairs but such gripping systems are hardly used in practice. Nevertheless, developing a sensor technology necessary for analyzing the workpiece to be gripped under such conditions is a major technical challenge. Using direct grip in such undefined situations a gripper cannot be expected to perform a reliable pick operation. Workpieces frequently have to be monitored again after the pick operation to make sure that they have been picked up safely. In addition to expensive sensor technology for workpiece analysis, the pick operation must also be monitored. So far the overall expense prevents an efficient use of grippers for this kind of application.

For workpieces which undergo further processing it does not make sense to reduce their order status by placing them into a box at random. A gripper placing workpieces into a box is generally used for reject goods as this undefined situation does not permit safe product placing. The workpiece falls from an undefined height onto other workpieces in the box which may cause workpiece damage.

Scenario 2: Workpieces With Unsorted Order Status On Plane Surface

In case a workpiece is isolated from bulk goods or presented to a gripper on a plane surface various sensors can analyze the position and orientation of the workpiece. As mentioned earlier, workpiece geometry determines so-called preferred workpiece orientation which already contributes information to workpiece monitoring. Monitoring situations which require more than just analyzing the position of single workpieces are a problem. This may be the case when workpieces overlap, e. g. if they are very close to each other or on top of each other.

overlapping workpieces

Such special cases are frequently complicated by product- or production-related exceptions. For product processing, for example, only workpieces of perfect quality are desired. Quality requirements are most diverse, e. g. surface roughness, form, or color, just to name a few. It is only the workpieces fulfilling these requirements which are to be handled. This quality assurance is not part of the handling task itself but a project of its own. It must be ensured that the handling system is not confronted with a workpiece of minor quality and thus not handling the wrong workpiece for nothing. Quality criteria must be clearly defined before starting to program an image processing or scanner software.

Workpieces may even happen to be in a position which is not suitable for pick operations at all, e. g. in case a workpiece can fall into a position where it hides suitable contact surfaces from the gripper`s operating elements.

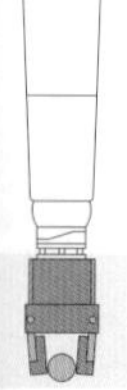

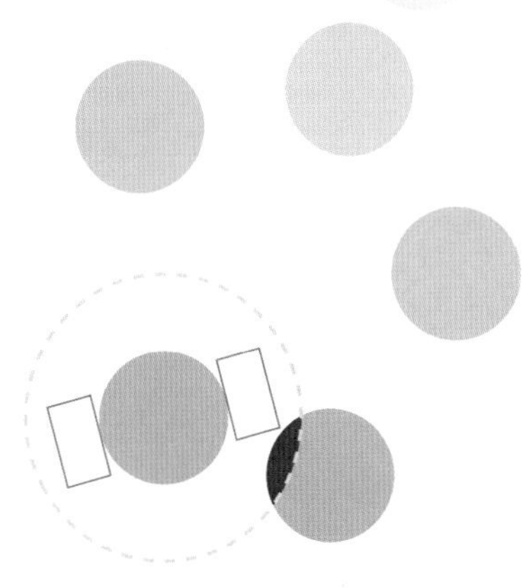

Figure 3.19 Interfering edges of workpieces in unsorted order status

Successful pick operations are dependent on the handling system`s degrees of freedom. The situation for workpiece compounds may again lead to collisions between gripper jaws and any workpieces which happen to be near the workpiece due for gripping.

If the gripping situation is monitored by sensors, the gripper can be positioned by the handling system to avoid collisions. According to workpiece proximity during preparation and the required "gripping zone" around the gripper it may occur that workpieces cannot be picked as prepared. These workpieces will have to remain in preparatory state for another try.

The workpieces which have not been gripped the first time because of their faulty degree of orientation or due to unsuitable gripping conditions, e. g. workpieces in danger to be damaged, can be prepared anew for the pick operation. This situation frequently occurs with small workpieces fed in great numbers.

Place operations of workpieces under such conditions run similar risks as described in the first scenario, workpieces may be damaged as well. If the workpiece is naturally stable, at least the order status can be maintained with the result that a pick operation for further processing is much easier.

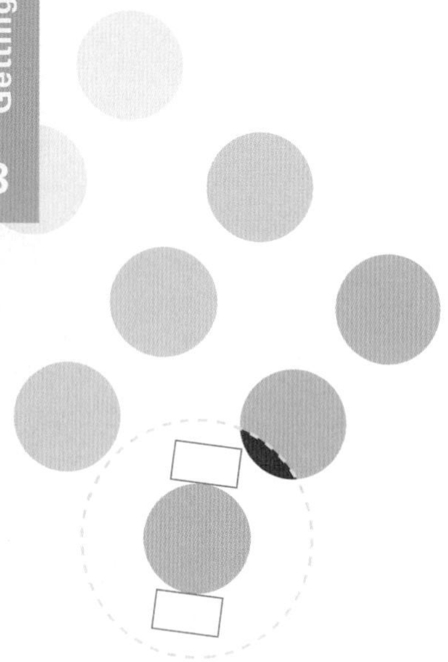

Figure 3.20 Interfering edges of workpieces in sorted order status

Scenario 3: Workpieces With Sorted Order Status

For a regular pick operation in industrial handling the workpiece is normally prepared in sorted order status. The workpieces` degree of orientation is largely maintained with the help of manufacturing technology in order to realize gripping without having to resort to expensive sensor technology. Careful planning is essential to avoid possible collisions of the gripper fingers with adjoining workpieces or unsuitable gripper housing.

Workpieces are frequently prepared on pallets for the pick operation. Construction engineers try to pack as many workpieces as close as possible on a pallet for maximum warehouse capacity. This objective often clashes with the need of maximum gripper flexibility for workpieces of various diameters. Figure 3.20 shows that the selection of an appropriate gripper not only depends on the workpiece itself but on how it is prepared on a pallet leaving the space necessary for the gripper jaws to pick it up safely.

Similar collision-prone situations occur when workpieces are fed into processing machines. Pick operations with chucks or similar make accessibility difficult. Pick operations with lathe chucks and short workpieces are a great challenge because the position of the lathe chuck jaws needs to be taken into account for the pick operation as well.

For a place operation the workpiece`s weight needs to be considered as this force may cause it to fall out of the gripper. Unwanted changes in workpiece position may occur if the gripper is opened before the workpiece can be safely clamped again.

Special Challenges For Grippers In Motion

More and more machines and component functions of production systems are directly linked to each other. This interlinkage demands continuous materials flow which possibly should exclude buffers as the latter will frequently change a workpiece`s degree of orientation and require additional investment resources. The three scenarios for pick operations as described above often occur in case of interlinked machines overlapping with workpieces in motion.

Pick operations for workpieces in motion can be distinguished as follows:

1. Pick operation without relative movement from gripper to workpiece **Vg ≠ Vw**

2. Pick operation with relative movement from gripper to workpiece **Vg = Vw**

Many handling systems already connect workpiece and gripper movement and convert workpiece movement into the respective gripper system of coordinates without any problem, i. e. synchronizing workpiece movement with robot movement.

Problems occasionally arise when workpieces are picked in motion, e. g. from a steadily moving conveyor, which may lead to positioning errors at the place station. Figure 3.20 illustrates the problem of a two-finger parallel jaw gripper trying to pick workpieces from different positions on the conveyor.

In the first picture of table 3.21 the workpiece moves with its contact surfaces, which are supposed to be touched by the jaws, in the same direction as the conveyor. The handling system positions the gripper above the workpiece and parallel to the movement direction of the conveyor and synchronizes it with the latter.

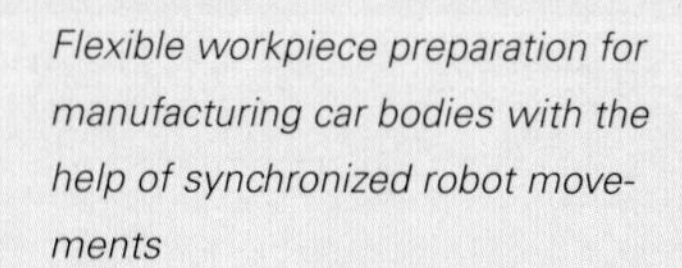

Flexible workpiece preparation for manufacturing car bodies with the help of synchronized robot movements

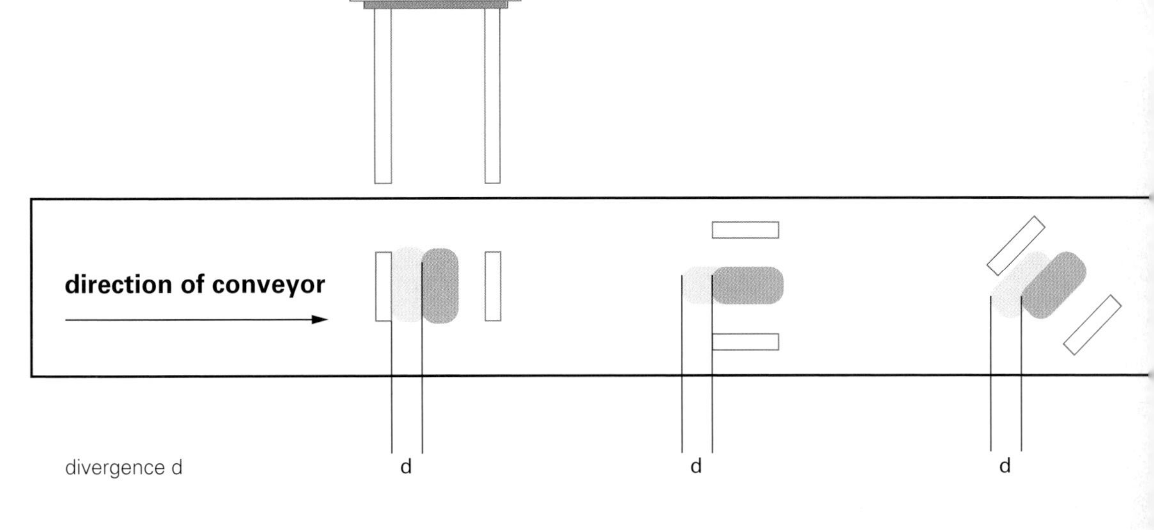

Figure 3.21 Workpiece divergence as a result of faulty synchronization during transport on conveyors

Synchronizing gripper and workpiece movement nearly equals the workpiece situation at rest. Therefore, workpieces cannot be misplaced during pick operations when the gripper closes with the gripper jaws reaching the workpiece at the same time. In case the gripper is not synchronized or positioned correctly in relation to the conveyor, a divergence between workpiece and gripper occurs. In a worst-case scenario this divergence may lead to a collision between gripper jaws and workpiece. If workpiece contact surfaces are aligned with the conveyor`s movement direction, it can be assumed for a two-finger parallel jaw gripper that workpiece positioning will not be influenced.

The second picture of figure 3.21 shows a workpiece with its contact surfaces relevant for the pick operation moving vertically to the direction of the conveyor. Synchronizing and positioning errors may lead to faulty positioning of the workpiece within the gripper as illustrated. This error is critical with regard to the subsequent place operation.

If the workpiece contact surfaces are situated diagonally in relation to the movement direction of the conveyor, velocity components along and diagonally to this direction are the consequence of the workpiece hitting the first gripper jaw. Thus the workpiece will not able to reach the correct position within the gripper. It is evident that accurate gripper positioning in relation to the workpiece is essential for successful pick operations.

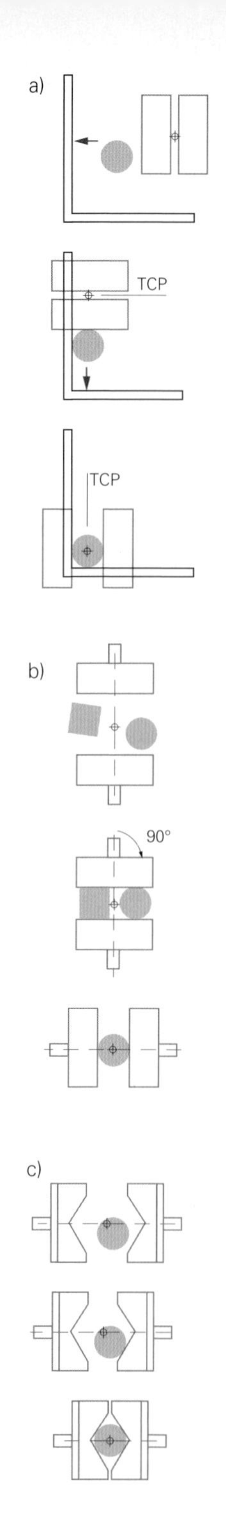

For applications requiring very high cycle times the workpiece is preferably picked up on the fly. This type of gripping is characterized by the effort to pick the workpiece up without letting the handling system move into a set position for the pick operation. The type of movement is generated by "overlapping" positions and is possible only if the workpiece has degrees of freedom along the movement direction.

As illustrated in figure 3.22, proper gripping strategies can be developed for picking workpieces up safely. By means of these strategies the workpiece can be well positioned within the gripper without having to resort to expensive sensor technology. In addition to using gripper movements to adjust workpieces for the pick operation, specially selected gripper jaws can help centering the workpiece.

This type of pick operation requires the workpiece to be positioned at a stop ring which supports positioning with the relative movement. The gripper jaws can be used as stop rings as well.

Gripping Accuracy Control

As detailed above precise presentation of the workpiece and accurate gripping during pick operations are essential for reliable place operations. Any errors in a pick operation can only be compensated by appropriate gripper or handling system sensors at a later stage. With smaller tolerances picking errors can be compensated by feed rails. Three reasons for faulty positioning of the workpiece are distinguished:

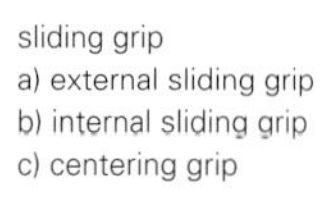

sliding grip
a) external sliding grip
b) internal sliding grip
c) centering grip

TCP = tool center point

Figure 3.22 Gripping strategies

1. faulty positioning of the workpiece before pick operation

2. faulty positioning of the gripper in relation to the workpiece (handling system error)

3. workpiece slipping within the gripper at gripper jaw closing or caused by faulty contact surface combination or gripping forces

Faulty positioning of the workpiece might be due to faulty synchronization of the gripper in relation to workpiece movement on a conveyor or workpiece support as described above. Other reasons could be faulty clamping devices or hazardous materials between clamping device and workpiece.

In any case it is important to pay attention to the degrees of freedom the workpiece has while being gripped, i .e. if the workpiece is still in the preparatory position when the gripper jaws close or if it is able to move within certain degrees of freedom. If the workpiece cannot be adjusted, faulty positioning of workpieces may cause premature wear or damage of gripper or handling system in the long run.

The same applies to faulty positioning of the gripper in relation to the workpiece. Integrating a mechanical collision and overload protection unit between gripper and handling system is one way to avoid strain or damage (see Chapter 4). This protection measure can be applied in case of workpiece tolerances leading to bracings.

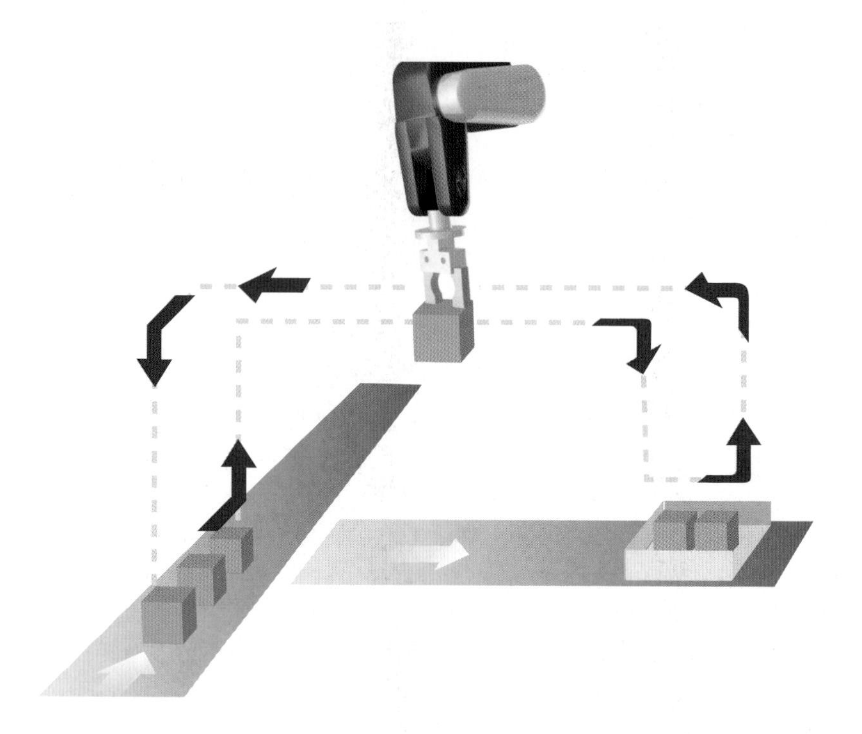

Figure 3.18 Direct linkage of workpiece input with workpiece output in a place position

Gripping Takes Time

Automated systems are either designed as self-supporting systems or with a direct link to existing production facilities. In this respect they are not directly integrated into the materials flow.

If direct linkage to pre- or post-operating facilities is selected, each workpiece fed into the handling process is due for further use or processing. The distance between pick- and place station is effected with one or several workpieces within a gripper.
Each workpiece delivered in the desired quality and correct order status is supposed to be moved. For these situations handling time per workpiece is essential. Kinetic devices with their characteristics in terms of workspace and velocity (see Chapter 4) are of great importance, opening and closing times of grippers also matter for cycle time.

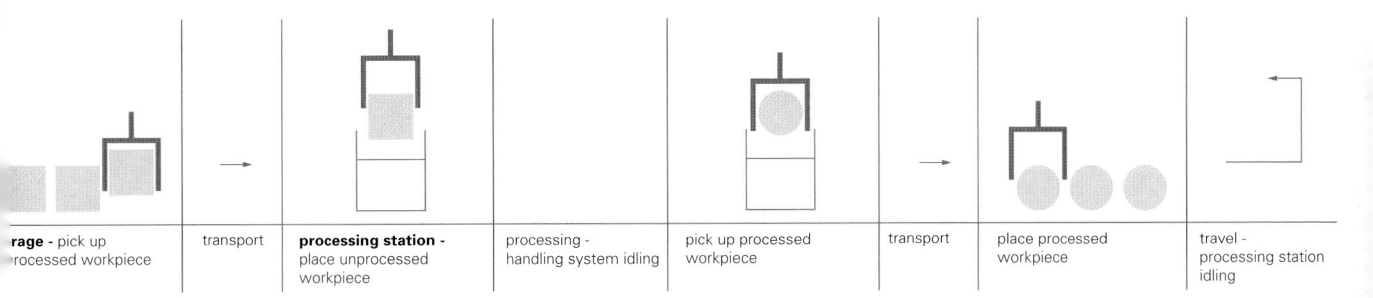

Figure 3.23 Materials flow for self-supporting system (without swivel unit)

Self-supporting systems or indirect linkages are decoupled from materials flow by buffers or magazines. The workpieces are taken from storage to a processing station where they are manipulated or reoriented and subsequently placed for storage again. The materials flow must be distinguished in as far as cycle times are selected with the result that the handling operation must wait for the processing machine. Swivel units are frequently employed to avoid idling during workpiece change in the processing machine.

Kinetic devices which carry the grippers are usually high-tech components which are much more expensive than the grippers. A six-axis robot, for example, costs far more than the gripper performing the pick operation. For most investments in automation profitability calculation includes handling output per minute, i. e. a robot will only pay if it is able to move a minimum number of workpieces per minute. These economic concerns leads from the principle of single gripping to the principle of multiple gripping to reduce cycle time per workpiece.

One benefit of single gripping is that gripper design does not need to be very sophisticated. A two-finger or three-finger gripper will be sufficient. Combined with a multi-axis robot the gripper can flexibly enclose workpieces because gripper positioning can be combined with most diverse workpiece orientations. In terms of workpiece positioning in the workspace the robot can make full use of its flexibility. Workpiece orientation is not important for reliable pick operations in this case.

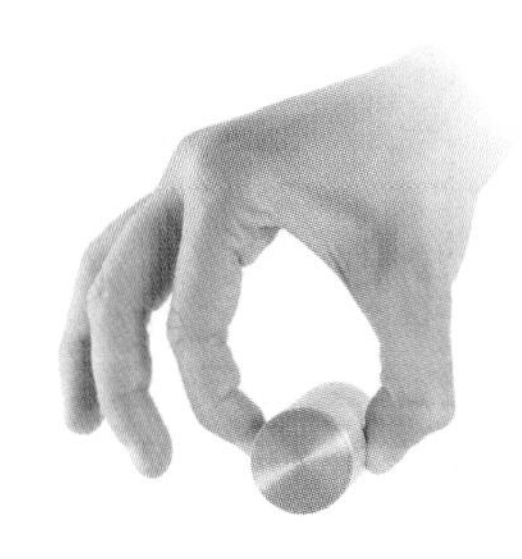

The same applies to place operations. Programming and putting systems into operation is streamlined. Interfering edges are reduced by compact gripper construction and low number of gripper fingers.

The disadvantage of the single gripper is that performance is directly coupled with cycle time of the axes in use. During one cycle from pick- to place position just one workpiece can be transported at a time. According to workpiece type and application, i. e. workspace and number of orientations, today`s robot kinematics operate at a regular cycle time between two and ten seconds per workpiece. Very few special kinetic devices designed for high-speed and low-weight handling tasks cope with cycle times far below these rates. Using parallel kinematics robots equipped with carbon fibre arms, for example, are able to reach up to 10g acceleration, which equals a ten-fold acceleration of the earth, and thus keep cycle times possibly under 0.5 seconds per workpiece.

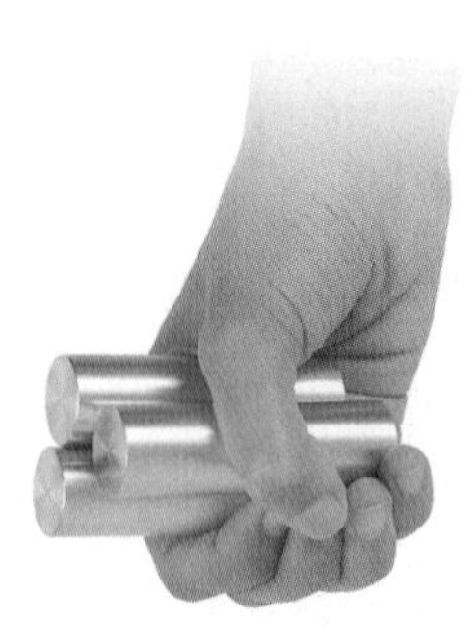

The **principle of multiple gripping** boosts the performance of a kinetic device. Figure 3.24 shows that a dual gripper already reduces cycle time by 50 percent compared to the single gripper at identical picking and placing times.

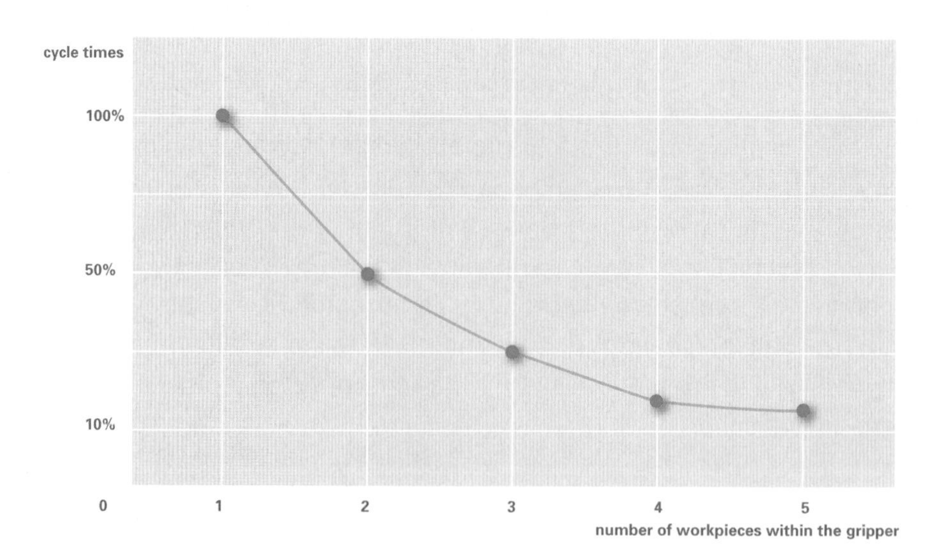

Figure 3.24 Efficiency increase by the principle of multiple gripping

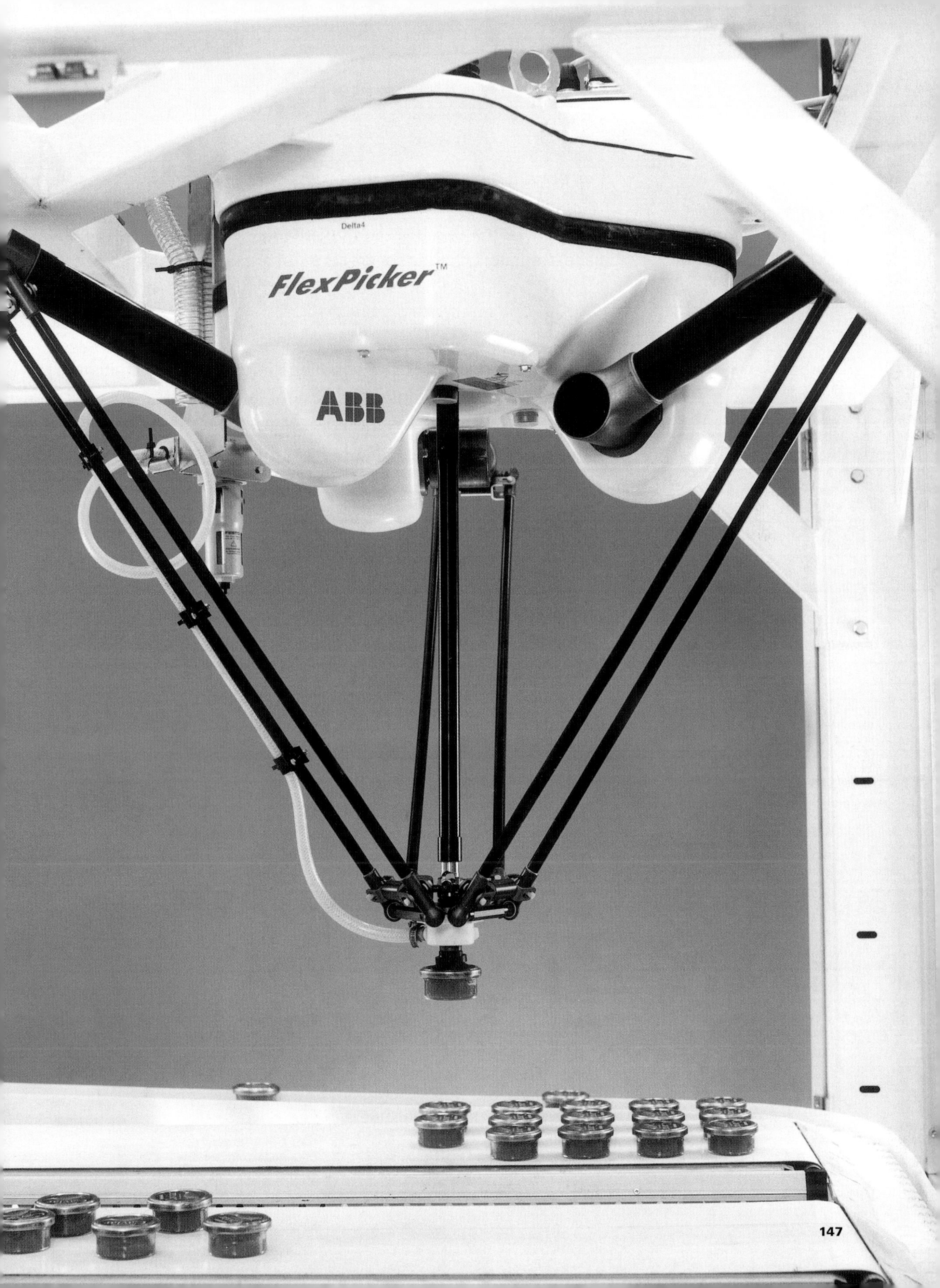
Delta4
FlexPicker™
ABB

This benefit pays because the number of kinetic devices (and their peripheries) is reduced and workspace saved. These cost savings, however, are partly offset by considerable expenses for pick- and place stations. Substantial preparation and implementation costs must be accounted for because the workpieces need to be aligned in groups for multiple gripping.

The process of aligning the workpiece in groups for a multiple gripper requires substantial investment into the periphery of kinetic devices. As illustrated the workpieces must first be aligned in a row with the appropriate distance to each other to be ready for the pick operation.

Sometimes the task is not limited to mere re-positioning of the workpieces. The alignment pattern for placing the workpieces may be different to the one at the pick station which makes multiple gripping more difficult. For many applications an option to change the distance between workpieces within the gripper works well. This changing option is frequently used for grippers in the packaging industry to take several workpieces at a time from a preparatory pick position to a different place position in a cardboard box.

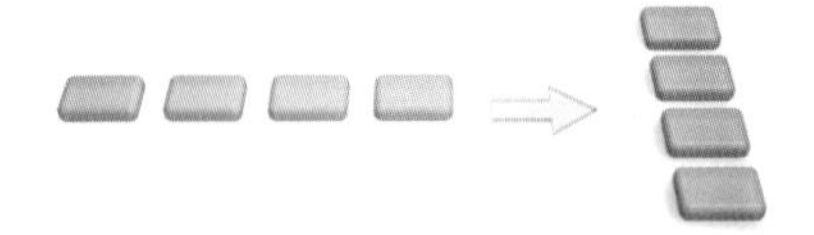

Exemplary alignment of workpieces before and after gripping

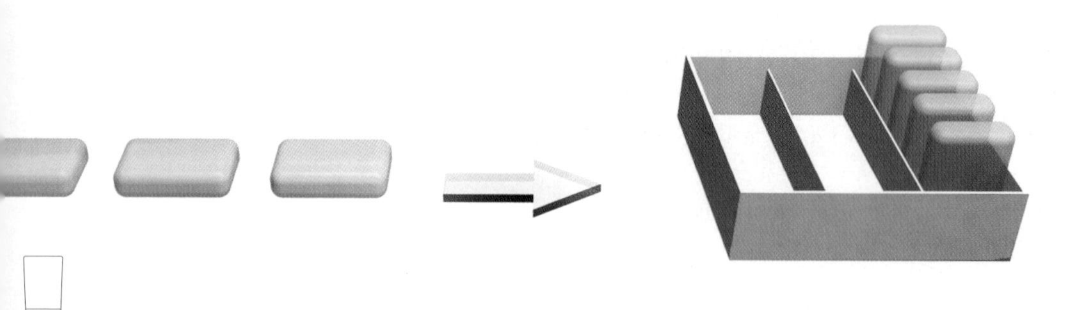

Workpieces which must be rotated into a vertical position make the handling task far more complicated

The workpieces in the example must be rotated 90 degree before they can be picked up by the multiple gripper.

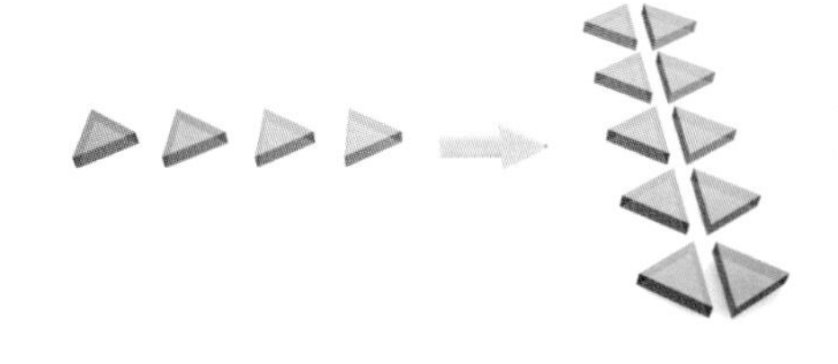

Alignment pattern for place operations requires rotating workpieces 90 degree

Such supposedly easy changes of workpiece orientation cannot be performed efficiently by multiple grippers in practice. The expenditure for process reliability and peripheral adjustment is substantial compared to the single gripper which often pays better despite higher cost for the handling device.

The principle of multiple gripping described so far is based on the assumption that all workpieces are picked up by the gripper at the same time and subsequently placed at the same time.

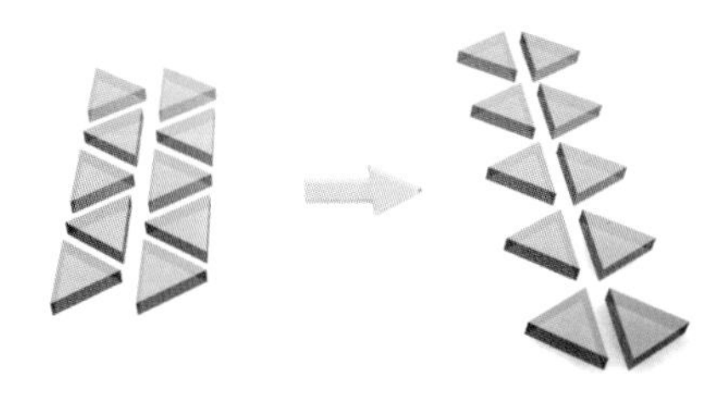

Task includes pick operation from a conveyor optimized for blend and place operation for packaging

The principle of stack grippers which has first been introduced at the AUTOMATICA 2004 in Munich, Germany, functions according to the single gripper concept but is able to store several workpieces within the gripper. Depending on workpiece dimension the gripper offers a saving capacity of minimum two workpieces and extremely short opening and closing times. For efficiency reasons, the time for picking a workpiece up and storing it within the stack gripper must be shorter than the time for the handling movement (transport) performed by the robot or axis system.

Several successive pick- and place operations performed by the stack gripper make cycle time shorter than single pick- and place operations according to the classic principle of single gripping. The stack gripper combines at least part of the benefits of the single gripper with those of the multiple gripper. Moreover, this gripper has a so-called dual stroke option. The latter reduces cycle time is reduced because the gripper is able to pick workpieces one after the other (in sequence) and release them all at the same time (parallel) at the place station.

Placing in stacks, 90 degree rotation of stack

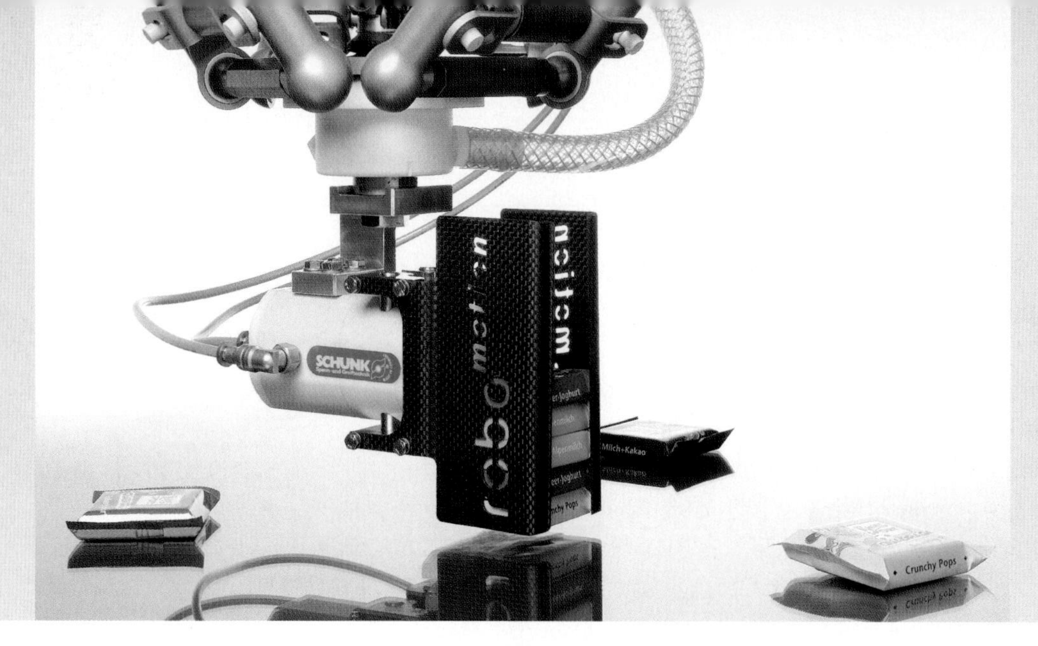

Workpieces stored within a stack gripper

In addition to placing them in stacks the workpieces can be placed in rows if the gripper is first turned into a horizontal position by a rotary unit or by the kinematics. Although this principle of stack gripping cannot be compared to the performance of a parallel gripper it does perform much better than a regular single gripper. The advantages of performance have been identified for parallel kinematics, also called delta kinematics, in tables 3.22 to 3.26.

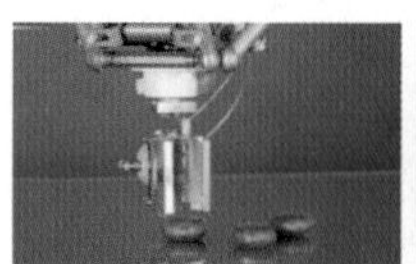

Pick operation of stack gripper

A higher workpiece weight is calculated for a multiple gripper because the number of workpieces stacked within the gripper increase the overall weight (table 3.22).
Compared to a regular single gripper a stack gripper can be expected to improve performance by nearly 20 percent if four workpieces are stored within the gripper (table 3.23).
The same stack gripper with a storing capacity of eight workpieces will increase performance by about 30 percent (table 3.24).
If the entire stack is placed at once performance can even be raised by 75 percent if the stack gripper holds four workpieces (table 3.25).
Pick- and place performance can be more than doubled by about 116 percent if the stack gripper can store eight workpieces and place them at the same time (table 3.26).

Delta kinematics single gripper	
single gripper (1x pick 1x place)	
ambient conditions	
width conveyor	1,000 mm
distance conveyor - place position	400 mm
average path	700 mm
gripper weight + 1x WP max.	500 g
result	
cycle	95.2ppm (parts per minute)

Table 3.22 Pick and place one workpiece (WP) with single grip (source: robomotion)

Pick with single grip and place in stacks

Delta kinematics multiple gripper 1	
quadruple gripper (4x pick 4x place)	
ambient conditions	
width conveyor	1,000 mm
distance conveyor - place position	400 mm
average path	700 mm
gripper weight + 4x WPs max.	1000 g
result	
cycle	114.2ppm (parts per minute)

Table 3.23 Pick and place workpieces (WPs) one by one: 20% performance increase compared to single grip

Delta kinematics multiple gripper 2	
octuple gripper (8x pick 8x place)	
ambient conditions	
width conveyor	1,000 mm
distance conveyor - place position	400 mm
average path	700 mm
gripper weight + 8x WPs max.	1,000 g
result	
cycle	123.6 ppm (parts per minute)

Table 3.24 Pick and place workpieces (WPs) one by one: 30% performance increase compared to single grip

Delta kinematics multiple gripper 3	
quadruple gripper (4x pick 1x place)	
Ambient conditions	
conveyor width	1,000 mm
distance conveyor - place position	400 mm
average path	700 mm
gripper weight + 4x WPs max.	1,000 g
result	
cycle	167.5 ppm (parts per minute)

Table 3.25 Pick workpieces (WPs) one by one and place them in one stack: 75% performance increase compared to single grip

Delta kinematics multiple gripper 4	
octuple gripper (8x pick 1x place)	
ambient conditions	
conveyor width	1,000 mm
distance conveyor - place position	400 mm
average path	700 mm
gripper weight + 8x WPs max.	1,000 g
result	
cycle	206.5 ppm (parts per minute)

Table 3.26 Pick workpieces (WPs) one by one and place them in one stack: 116% performance increase compared to single grip

3.5 Safe Gripping

Losing our grip on a coffee-cup or accidentally letting a jar of mixed pickles slip from our hands in the supermarket does not do great harm. Losing the grip on workpieces during handling, however, may lead to major financial damage. For example, a workpiece accidentally lost in a processing machine may cause serious mechanical defect after re-start. Just imagine a workpiece within the gripper of a robot rotating with an action radius of three feet at full speed turning into a kind of projectile, even more dangerous at a robot payload up to 1,100 pounds. High-grade workpieces require maximum protection against loss or damage, too.

Risk of workpiece loss or damage is evaluated with the help of the Failure Mode Effect Analysis (FMEA) which has become part and parcel of a methodical handling task approach. Risk evaluation is a future-oriented method for analyzing potential hazards and the probability of such hazards. Beyond mere damage repair this method is a significant step towards far-sighted and safe gripper design and construction.

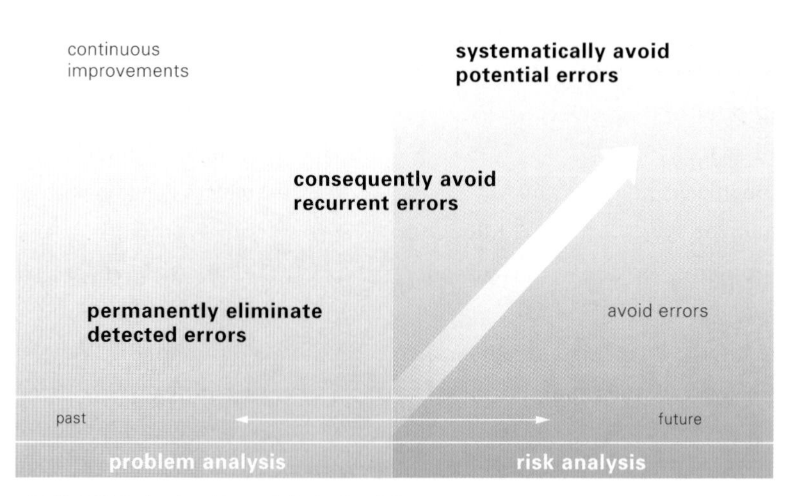

Grafik 3.25 Evaluating risks and eliminating defects

Potential hazards due to workpiece loss are evaluated, e. g. on a scale of 0 to 10. In the next step the probability of occurrence for these hazards can be assessed. The probability of detecting such a hazard is evaluated as well. If you then multiply the potential hazard with both its probability of occurrence and its probability of detection you get the risk factor for the overall risk assessment.

The example for a risk analysis in table 3.27 is based on the risk of workpiece loss by a pneumatic gripper, i. e. the workpiece sliding within the gripper. Gripping force is substantially reduced by a drop in pressure of 2 bar from a standard pressure of 6 bar. Analyzing the situation for the entire system means looking at its interfaces in relation to all other components, not just to the gripper. For the entire system a workpiece loss is evaluated as highly critical, e. g. risk of personal injury or material damage. The risk of faulty positioning of the workpiece within the gripper is rated low, e. g. concentric gripper jaws do not let the workpiece slide but into one direction (z). Nevertheless, the probability of occurrence for sliding is rated higher than for total workpiece loss because the workpiece would first have to slide out of the gripper jaws before it could get lost. The probability of detecting a pressure drop to 2 bar is rated relatively high due to the pressure control sensor installed which warns the operator. As the operator may fail to notice this warning the risk analysis result is still rated at 4.

description of hazard (1-10)	potential hazard (1-10)	probability of occur-rence (1-10)	probability of detection (1-10)	risk factor (1-1000)
loss of workpiece at pressure drop to 2 bar	8	4	4	128
workpiece sliding at pressure drop to 2 bar	2	8	4	64

Table 3.27 Comprehensive risk analysis

Table 3.27 shows that a comprehensive risk analysis will only be feasible if the hazards are known for the entire system the gripper operates in. A workpiece sliding within the gripper may result in serious damage for the periphery of an automated solution and the handling unit as well as for the robot itself. Collisions with a work-piece out of place within the gripper are highly probable if space for the pick operation is limited. Table 3.27 is based on the assumption that a workpiece which is not properly held within the gripper can still be placed without collision.

This plain example shows that teamwork is indispensable for risk evaluation. Apart from know-how about the braking effect on a workpiece within the gripper it is essential to be aware of the kinematics in case of an emergency stop. The same applies for robot and control know-how.

The comprehensive risk analysis has been developed for teamwork and is based on all staff members contributing to the evaluation on the basis of their special know-how. This is the only way to get anywhere near to reliable assessment of potential hazards and risks. So far standards (DIN1672-2:2004) have been set which require robot producers to evaluate certain types of risks such as hygiene-related risks.

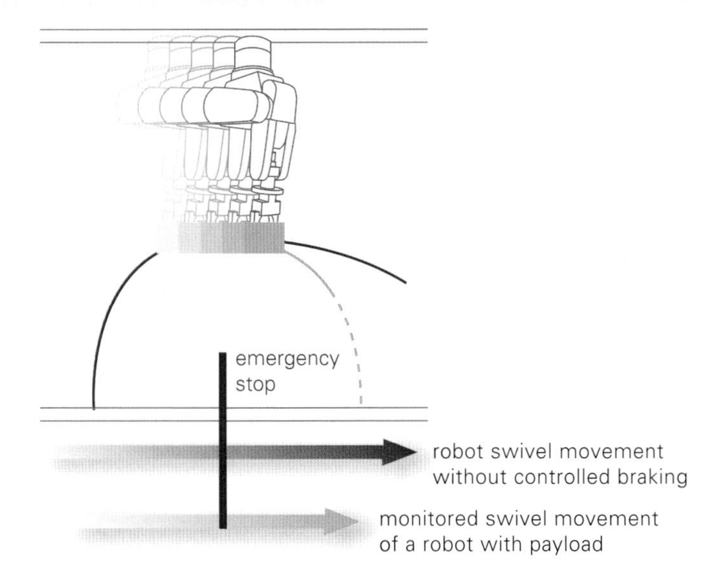

Figure 3.26 Robots behaving differently in emergency stop situation

Two emergency stop situations at the risk of losing grip on the workpiece are explained in the following. Any situations which may occur due to workpiece range, workpiece tolerances, or friction tolerances, are neglected. Workpiece loss mainly occurs during **emergency stop** and **power failure**.

In case of an emergency stop a sudden brake is put on the robot arm or the kinematics. This brake leads to forces of inertia on the workpiece which are not taken into account in the regular gripping force calculation and which may lead to workpiece loss at the worst.

The workpiece`s center of gravity is the ideal gripping point in order to achieve minimum torque during workpiece movement. It depends on workpiece geometry whether this can be achieved or not.

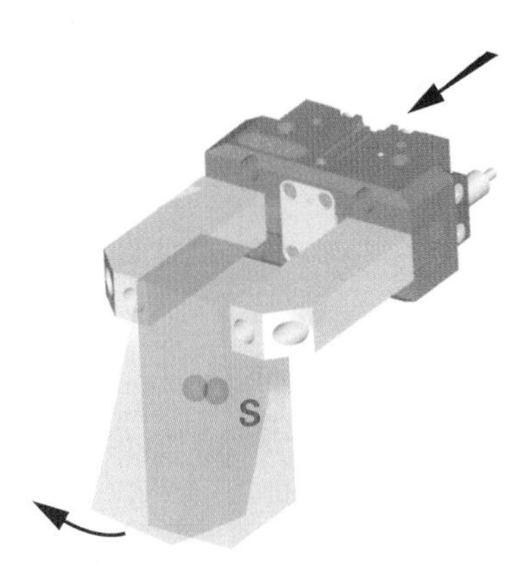

Figure 3.27 Braking movement of a robot with an unfavorable workpiece geometry within the gripper

If the energy supply for the gripper is reduced or completely cut off the workpiece is most probably lost if the respective provisions for gripping force maintenance have failed to be made beforehand.

spring-supported gripper

The use of spring elements which press the piston of pneumatic grippers into the closing position is one possibility for maintaining the gripping force. The spring supports the closing force with air pressure during normal operation and closes the gripper.

For gripping force maintenance tension springs or compression springs can be used. Housing restrictions, e. g. for compact grippers, are often responsible for the fact that only part of the maximum gripping force can be secured.

This is an acceptable compromise because it is sufficient to safeguard the static load of the gripper, i. e. holding the workpiece within the gripper while it does not move, as a loss of energy usually leads to an emergency stop. Spring elements maintaining the gripping force are also used for securing workpieces during longer standstills, e. g. over weekends, as this reduces starting times.

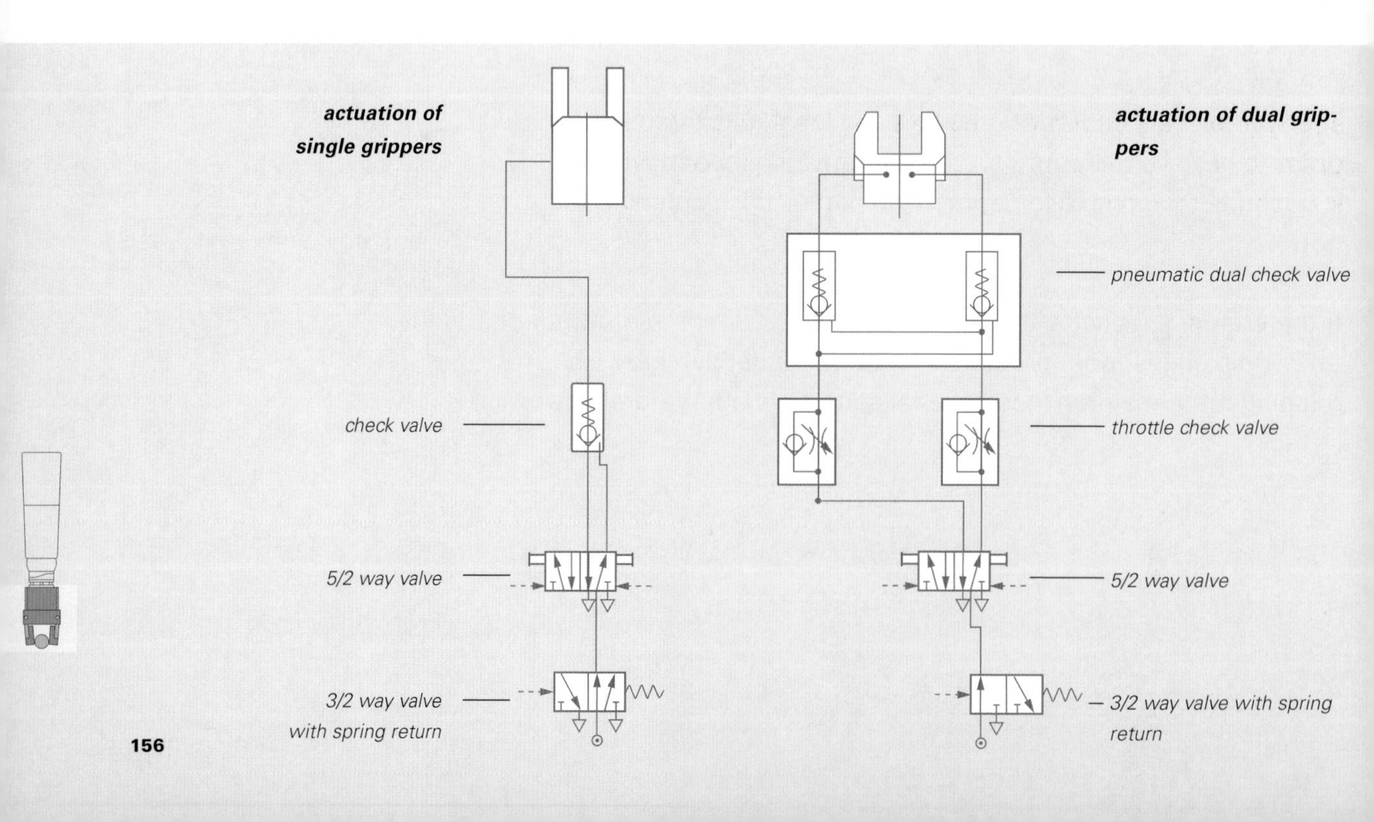

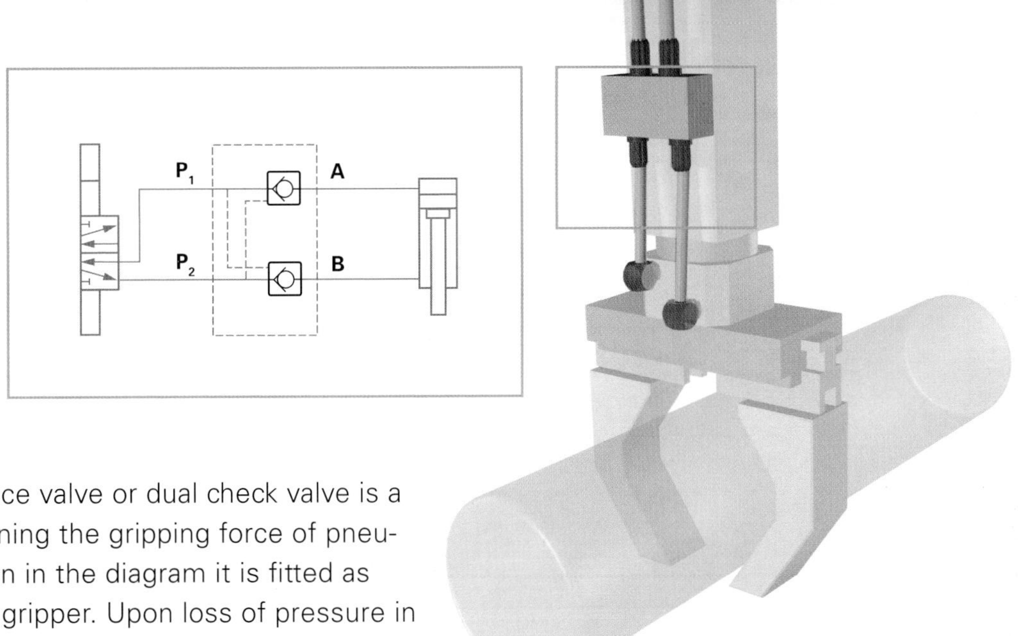

The pressure maintenance valve or dual check valve is a safe method for maintaining the gripping force of pneumatic grippers. As shown in the diagram it is fitted as close as possible to the gripper. Upon loss of pressure in the supply the dual check valve closes to prevent air leakage from the gripper cylinder.

principle of function for the dual check valve

Dual check valves are employed for pneumatic as well as hydraulic grippers to maintain the gripping force. Their advantage is that they require no additional hoses within the gripper´s immediate workspace and permit flexible fitting.

The valve must be fitted to the gripper the closest possible to enable fast reaction upon loss of air and to keep the part of tube supply which could be affected by a leakage as small as possible.

The dual check valve is used if gripping force maintenance by spring force is not possible at all or only at unreasonably high expense. This is the case when the housing necessary for the springs safeguarding high clamping forces is not sufficient.

3.6 Grippers as a Source of Information

Copying human abilities such as sense of touch and visual perception of the gripping action the gripper not only to manipulate the workpiece but to analyze it. Both basic functions are integrated into grippers.

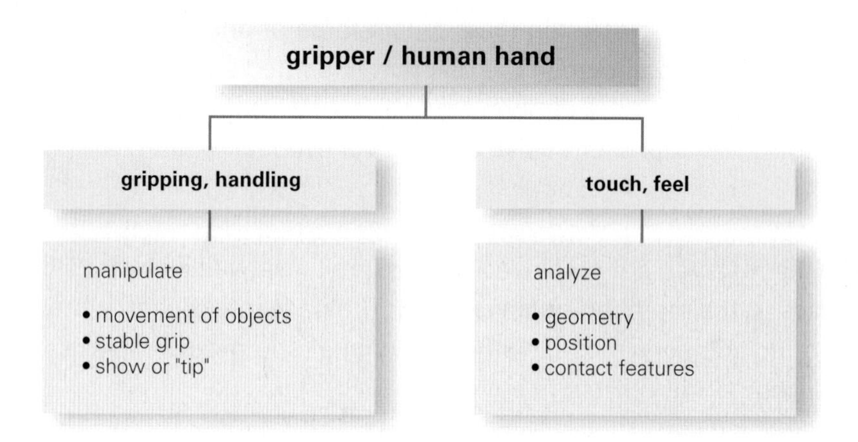

The human hand is both a sensor and handling tool in one.

Human sensors work through the sense of touch or, in other words, are tactile sensors. In addition to tactile sensoring humans can visually monitor their handling operations which is a non-tactile function. In automation sensors are generally categorized as tactile and non-tactile. Tactile sensors require force or torque acting directly on them to be able to record anything. Non-tactile sensors work at a distance between sensor and object of monitoring.

Table 3.28 includes examples of tactile and non-tactile sensors in automation technology and how they are categorized. The tactile sensors are subdivided into sensors which are able to measure a certain force or torque, and sensors which merely trigger a control mechanism.

sensors in automation technology			
tactile	**non-tactile**		
force/torque	**video-visual**	**ultrasonic**	**other**
• multicomponent force/ torque sensor • gripping force measure • active wing/blade gripper • RCC • IRCC	• linear sensor • image processing (binary, gray scale value) • 3 D stereo imageprocess-ing • image processing with active illumination	• proximity switch • sonic barrier • distance measuring • scanner • acoustic correlation sensor	• microwave • pneumatic • radioactive • chemical
tactile	**visual**	**inductive, capacitive, magnetic and piezoelectric**	
• switch • distance measuring • touch line • touch matrix • flat-top switch • slip sensor	• light barriers • reflection light master • distance measuring • 2 D scanner • 3 D scanner • light stripe sensor • visual correlation sensor	• proximity switch • distance measuring • welding seam tracking • vibration analysis	

Table 3.28 Overview on sensors in automation technology

The non-tactile sensors include examples of visual and video-visual sensors. This group of non-tactile sensors works with light as measuring medium. Physical principles employed are ultrasound, induction, capacity, magnetism and piezoelectric effects. Microwave and pneumatic sensors are used in automation as well. Other methods are being developed so that more and more applications can be realized.

For mechanical gripping it is important to know that according to the information required the sensors are integrated into the gripper. Direct contact with the workpiece is established which easily permits tactile measuring, e. g. when accurate information on gripping force is required.

As shown in table 3.26 both non-tactile and tactile sensors can be further classified into **switching sensors** and **measuring sensors**.

Switching sensors are mostly used for determining if a particular position has been reached. This type of gripper monitoring has been used for over ten years and is standard for mechanical grippers. Process reliability of automated production systems could hardly be achieved without sensors. Cycle times of handling processes can be optimized by sensors instead of programming waiting periods for gripper opening.

Proximity switches, reed or magnetic switches are often empoyed as switching sensors.

magnetic switch

Switching sensors can monitor the respective gripper status, i. e. monitoring the positions as follows:

- gripper open
- gripper closed
- gripper closed/open for internal grip

Iinductive proximity switch

For the first two gripper positions it is clear where the sensors must be fitted. These two positions of the operating elements are usually final positions. The third position (gripper closed/open) is more difficult to monitor with switching sensors. Switching times or sensor fitting may vary according to workpiece tolerances or different workpiece dimensions.

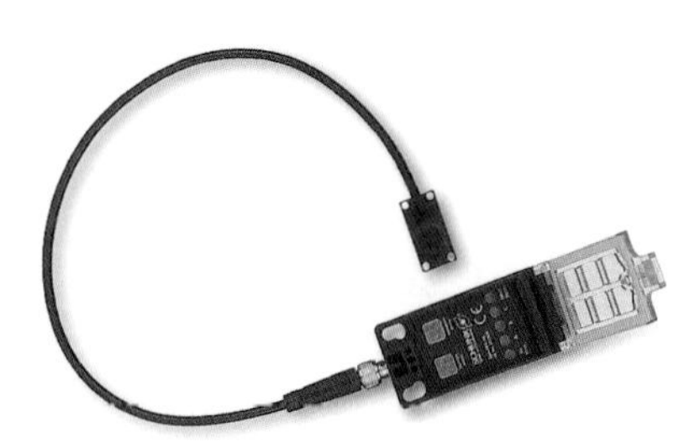

flexible position sensor

Nevertheless, there are ways to gain information on the respective gripper status for all three positions.

It may be sufficient to just check both final positions of a gripper`s operating elements, e. g. with two proximity switches. When the operating elements leave the position "gripper open" and do not reach the position "gripper closed" the workpiece is recognized as gripped. The gripper is unlikely to stop between its two final positions.

If the position of the operating elements is utilized for the information "workpiece gripped" the sensor must be exactly adjusted for this very position and for the relevant workpiece. This adjustment is done while the workpiece is within the gripper. The sensor can be pushed in a slide bore up to point where the workpiece is being gripped and thus be adjusted. Another option is the use of banners at the operating elements which can be adjusted accordingly.

By adding a third sensor both final positions and the interim position "workpiece gripped" can be monitored. Switch positions can be pre-set for defined areas as shown in the diagram.

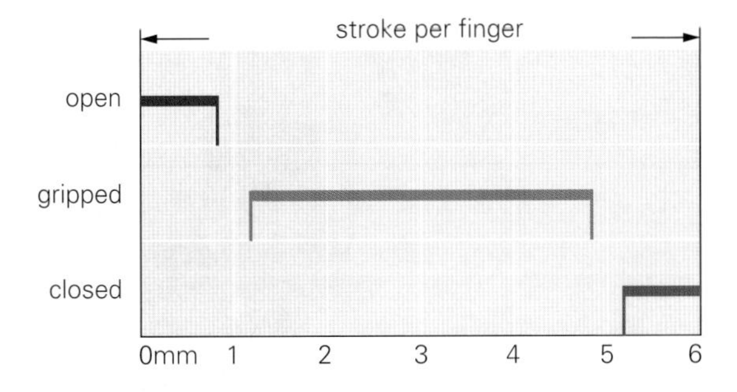

Figure 3.28 Pre-set areas where the proximity switch releases a signal

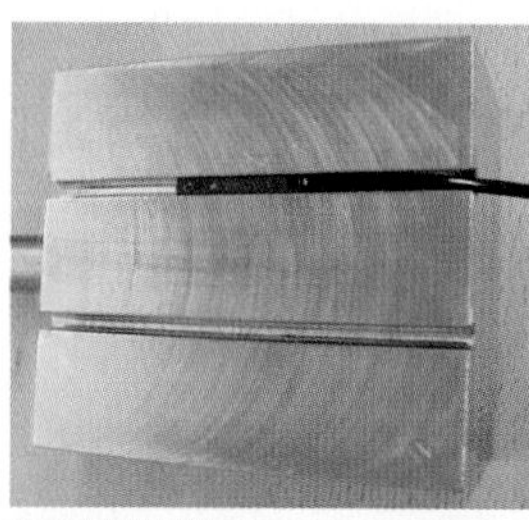

measuring the piston position by magnetic sensors in the tongue

For workpieces with especially great tolerances two sensors can be used to limit the space for gripping. The first sensor measures the operating elements entering this space while the gripper is being closed. While "crossing" the sensor a flank is measured. If the second sensor at the end of the gripping task area does not receive any signal it can be assumed that the workpiece has been picked up. If the second sensor is activated at the end of the gripping tolerance the pick operation has failed.

Hall effect sensors which are activated by magnets can be used for a broad range of applications as well. A magnet which is fitted to the piston of a pneumatic cylinder activates the sensor. Thus more than two switching points are available because the sensors can be fitted to the drive cylinder over the entire length of the piston movement.

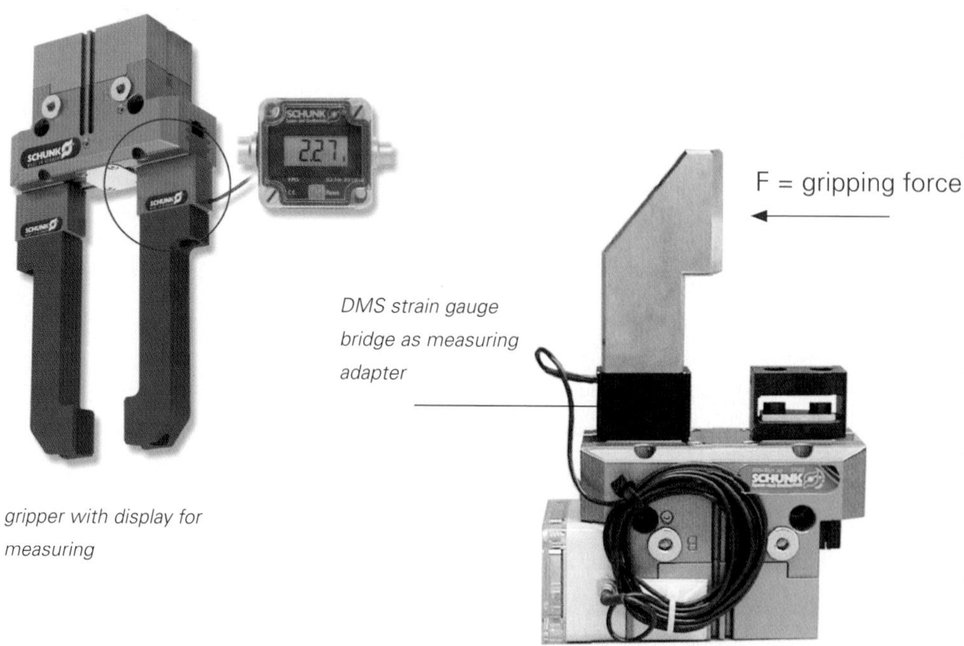

Figure 3.29 Force measuring adapter: monitoring gripping forces at the gripper fingers

Measuring sensors are normally used for measuring workpieces, for positioning the gripper stroke, or for controlling the gripping force. These measuring tasks may be of interest for pick- and place operations as well as for workpiece transport. In addition sensors can be fitted to the gripper flange for the moving unit (see Chapter 4).

The gripper fingers can be utilized as sensors for force induction by adhesive sensors which are fitted to the gripper fingers.
This so-called force measuring adapter can record and analyze gripping forces with the help of a DMS strain gauge bridge.
The force measuring bridge is mounted between finger and gripper kinematics for this purpose.

The result of this tactile measuring sensors is a so-called dynamic force measurement parallel to the gripping process. This measurement is visualized by a software which directly converts it into a diagram, e. g. for a 100 percent gripping control protocol. As a result workpieces can be classified according to the force reached. With the protocol data preventive gripper maintenance is achieved.

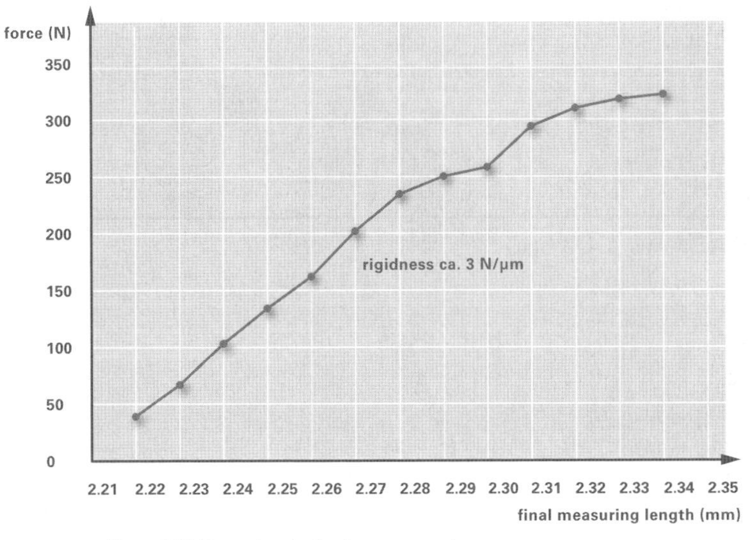

Figure 3.30 Dynamic gripping force measuring

SCHUNK
Spann- und Greiftechnik

A pneumatic drive is not sufficient if sensoring as well as active gripper control is necessary. For complex gripping tasks depending on active gripper control electrically driven grippers have become a clear trend over the past years.

Servoelectric drives are favored because they are highly flexible (programmable). Simple integration into bus systems and networks used in automation is a major benefit permitting online control of the gripper. Nevertheless, the use of electric grippers is still too expensive for many applications.

One example of an electric gripper with the respective sensors is the gripper with DC current motor shown here.

The gripper includes a spindle drive with magnetic brake and an incremental encoder for determining position and angle velocity. Precise guidance in combination with the ball revolution spindle permits an especially accurate and sensitive gripping force control.

Repeatability is as accurate as 0.05mm for the gripper fingers at a gripper stroke of up to 70mm. At closing of the gripper fingers a velocity of 82mm per second is achieved equalling a closing time of 0.85 seconds. Maximum gripping force is 200N.

Evaluation unit with circuit board

The finger grippers already mentioned require even more sensitive sensors. The fingers of the Barret Hand are based on DMS strain gauge technology which permits them to have a sense of touch. The evaluation unit which converts sensor data into positioning orders for the drives is also integrated into the hand in accordance with the modular concept. The figures show the sensors in the fingers and the evaluation unit with the circuit board in the flange of the hand.

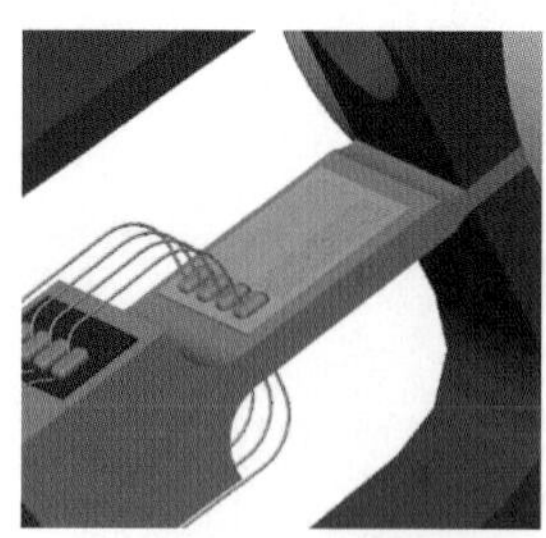

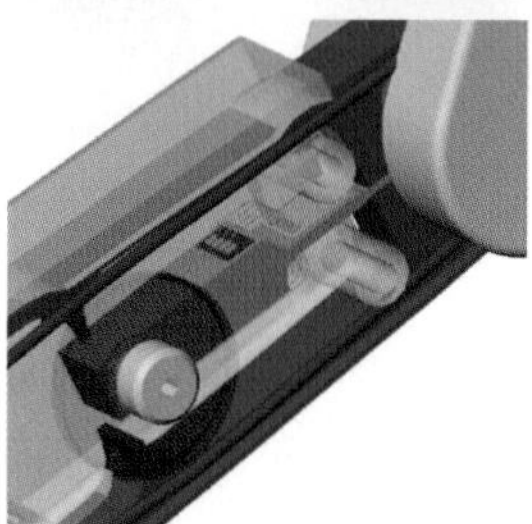

CAD image: sensors in the finger of the Barret Hand

A very special kind of sensor technology has been developed for the following gripper. This small parts gripper has an electric drive and can be equipped with up to 16 sensors. The construction integrates the entire control electronics into the gripper as well. The sensors can measure temperature, force, position, and even workpiece conductivity.

Current efforts in sensor integration signal a future trend for the following gripper attributes which will be available within the next few years:

- adaptability
- sense of touch
- visual acuity ("eyesight")
- interactivity ("acting and reacting")

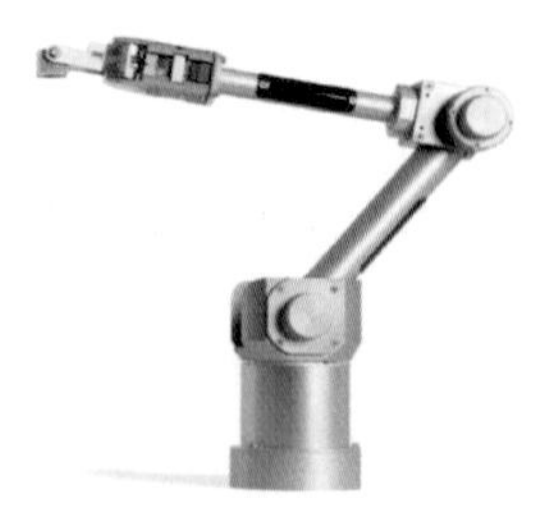

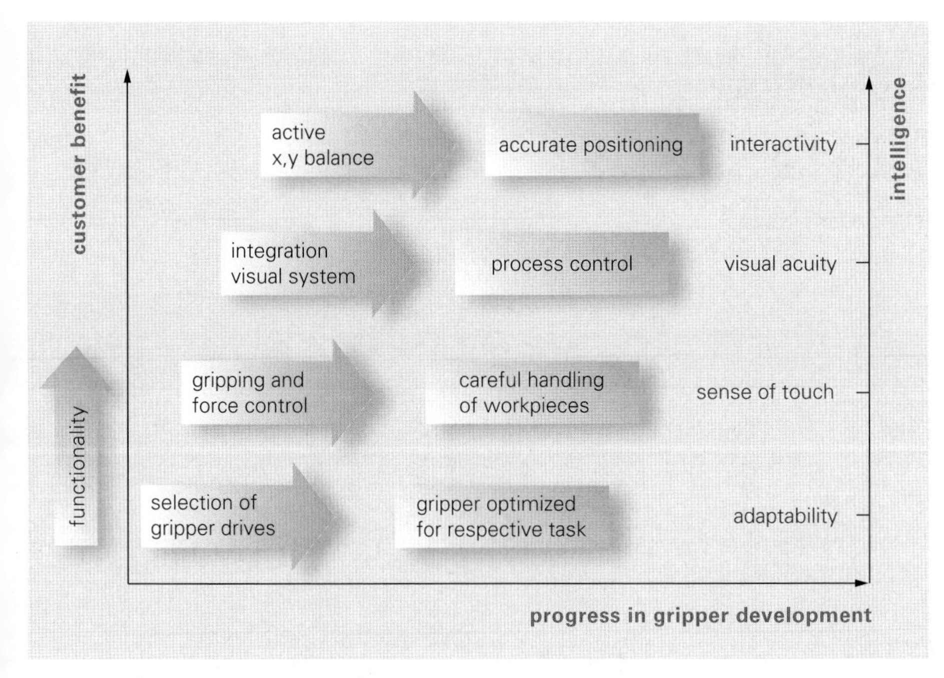

Figure 3.31 Gripper development trend for sensors

In short, sensors in combination with the respective drive and control technology make the gripper more intelligent. Gripper drives are expected to be available for individual selection according to the respective task. Thus the gripper can be fully adapted in terms of gripping radius , gripping control or gripping force. Gripping force control ensures careful and gentle workpiece handling. Additional functions such as workpiece measuring can be performed. Integrating image processing systems into or onto the gripper permits control on the gripping and handling process. An active compensation can be performed during the final step, such as a most accurate adjustment of the gripper jaws, which is very useful for workpiece insertion.

By all means today`s and tomorrow`s grippers are far beyond the VDI definition as described at the beginning of this chapter.
The Clamp and the Release function were at the center of task definition, current applications are much more challenging.

4 Movement Adds Value

After a close look at the pick operation in Chapter 3, the movement of the workpiece is now at the center of our attention. Dynamics treat forces as the cause of kinetic processes. The latter not only influence gripper design to a great extent but also determine the choice of the device for generating movement. The appropriate kinetic device must be selected, and the weights of both gripper and workpiece need to be taken into account as notable forces.

The profitability of an automated solution depends on velocity and cycle time as well. The output per hour is the determining factor for plant productivity and, in consequence, for economic efficiency.

plant productivity = output / time

The amortization of a machine is calculated on the basis of its ordinary service life to assess the return of such an investment. In terms of economic efficiency, an automated solution is usually compared to manual labor.

Nevertheless, this calculation fails to consider important aspects of safety, quality, or hygiene, as arguments in favor of automation. Whoever is able to value these aspects in terms of economic efficiency, will have sufficient stamina to open up new applications in automation and be patient enough to master inherent technical challenges.

As pointed out in Chapter 3, a pick operation performed by a gripper depends on various factors. Similar criteria influence the movement of a workpiece, not to forget the characteristics of the handling device. The process of moving workpieces with the respective gripper system, the kinetic device, and the place operation, need to be analyzed in detail.

Ambient conditions of the kinetic task	Workpiece/gripper combination features	Kinetic features
foreign material	form	kinetic measuring
economic efficiency requirements	dimension	kinetic form
workpiece variety	tolerance	degrees of freedom (dof)
temperature	mass	velocity
energy supply	moment of inertia	acceleration
installation options	center of gravity	precision
cleanroom	surface	repeatability
hygiene	consistency	risks of collision
maintenance-free		
safety provisions		
monitor process parameters		

Table 4.1 Essential features of the kinetic task

The ambient conditions of production are basically the same for kinetic devices as for the gripping process. The challenges which automation components are faced with, such as harsh environments, strict hygiene or cleanroom requirements, are also the same.

Other criteria, such as weight and dimension, which were formerly defined by the workpiece alone, are now defined by the workpiece/gripper combination. Forces caused by dynamics, which may lead to the loss of a workpiece, need to be taken into account when selecting the appropriate workpiece/gripper combination.

In this respect, kinetic features are just as important for the gripping process as for selecting the appropriate kinematics. Kinetic features in automation technology can be compared to those in sports medicine because velocity, acceleration, and precision of movement, are equally valued.

4.1 Kinetic Effects on Workpieces

Workpiece features have already been analyzed in Chapter 3 in terms of their relevance for gripping. When workpieces are analyzed in terms of their movement, the focus is set on mass distribution and workpiece consistency. Additional forces arise from the movement, which need to be compensated by the gripper. Such forces resulting from the movement can be divided into

1. forces of inertia
2. forces of process

installation options	direction of acceleration	force / required gripping force per gripper finger	
		$F_G = m(a_z+g)\frac{\sin\frac{\alpha}{2}}{2\mu}S$	
		$F_{G;z} = mg\frac{\sin\frac{\alpha}{2}}{2\mu}S$	$F_{G;x} = ma_x\frac{\tan\frac{\alpha}{2}}{2\mu}S$
		$F_{G;z} = mg\frac{\sin\frac{\alpha}{2}}{2\mu}S$	$F_{G;y} = ma_y S$
		$F_G = m(a_z+g)\frac{\tan\frac{\alpha}{2}}{2}S$	
		$F_G = m\left(a_x+g\frac{\tan\frac{\alpha}{2}}{2}\right)S$	
		$F_{G,z} = mg\frac{\tan\frac{\alpha}{2}}{2}S$	$F_{G;y} = ma_y\frac{\sin\frac{\alpha}{2}}{2\mu}S$
		$F_G = m(a_z+g)S$	
		$F_G = m\left(g+a_x\frac{\tan\frac{\alpha}{2}}{2}\right)S$	
		$F_{G,z} = mgS$	$F_{G;y} = ma\frac{\sin\frac{\alpha}{2}}{2\mu}S$

Table 4.2 Kinetic effects on the required gripping force per gripper finger

The following symbols are important for calculations:

Relevant symbols and their meaning			
a	acceleration	k	correction factor
a_R	radial acceleration overall rotary specs	l, L	lengths of links
		M	moment
a_t	tangential acceleration	m	mass
a_{NA}	emergency stop acceleration	p	normal pressure
a_z	central acceleration	P_O	over-pressure
A	plane	P_U	under-pressure
B	magnetic induction	r	radius
D	diameter	s	distance
E	elasticity module	S	security factor
F	force	t	time
F_C	Coriolis force	v	velocity, translation
F_G	gripping force	α	jaw opening angle
F_H	force to lift	β	auxiliary angle
F_{NA}	emergency stop force	φ	friction angle
F_S	force to fall	ς	opening angle
F_R	resulting force	μ	rotary angle
F_V	force to displace	μ	friction value
F_Z	centrifugal force	ω	permeability
G	weight	$\dot{\omega}$	angle velocity, rotary velocity
g	acceleration of the earth		

Table 4.3 Relevant symbols and their meaning

Effects of The Forces of Inertia

The forces of inertia result from acceleration of the workpiece`s mass.

$F = ma$

F = force [N], m = mass of the workpiece [kg], a = acceleration [m/s^2]

These forces must overlap the forces which result from the acceleration of the earth, in order to calculate the force required for force-fit gripping. Profound knowledge of the movements performed by the gripper permits a competent decision on the gripper construction. The following overview includes the necessary steps for analyzing workpiece kinetics and calculating the corresponding gripping force.

The following forces result from the so-called dead weight of the workpiece as well as from the translatory and rotatory movements of the gripper and the workpiece. All the forces need to be transmitted by the operating elements of the gripper.

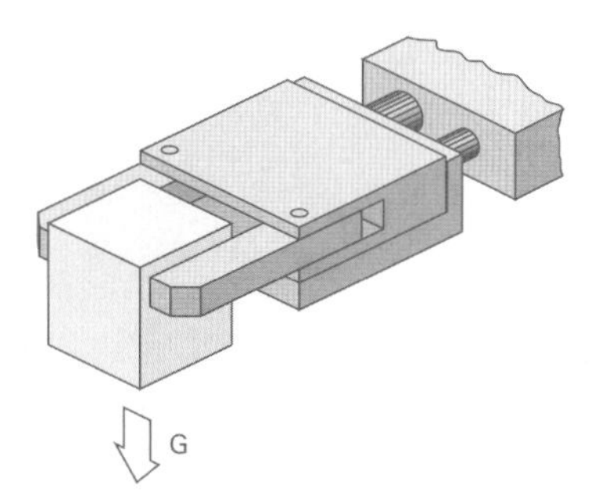

Figure 4.1 Static gripping: gripping force must compensate weight

The weight G is generated by the acceleration of the earth and acts as a force of the workpiece, directed towards the center of the earth at any moment of the handling process. Therefore, the weight is defined for the entire handling process in terms of size and direction. All other forces of inertia need to be calculated in relation to their size and direction for the respective path and acceleration.

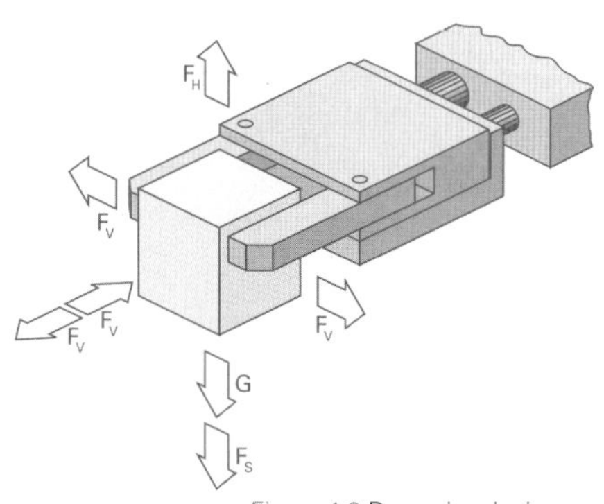

Figure 4.2 Dynamic gripping: forces of inertia must be compensated by translation

If the gripper is to perform a translatory movement with the workpiece, the resulting force FR must be described according to the direction of this movement. If the effective lines of a workpiece for lifting and falling are exactly parallel to the weight, the result is as follows:

Force to lift

$$F_H = mg(1 + \frac{a}{g})$$

Force to fall

$$F_S = mg(1 - \frac{a}{g})$$

Within some handling systems, dynamic robots can reach an acceleration up to 100 m/s^2. A modern industrial robot reaches a standard acceleration of about 20 m/s^2.

A force acting horizontally, i. e. perpendicular to the weight G, is calculated and described as the so-called force to displace.

Force to displace

The force acting on the workpiece, while taking it from one point to another, is compensated by the operating elements in the same way as during form-fit gripping. As illustrated, this occurs when a purely horizontal movement is carried out perpendicular to the operating elements. The importance of form-fit gripping (see Chapter 3) for the required gripping force becomes evident at this point.

The resulting force for the horizontal translation of a workpiece is calculated as follows:

$$F_R = \sqrt{G^2 + F_V^2}$$

with the direction angle

$$\rho = \arctan \frac{F_V}{G}$$

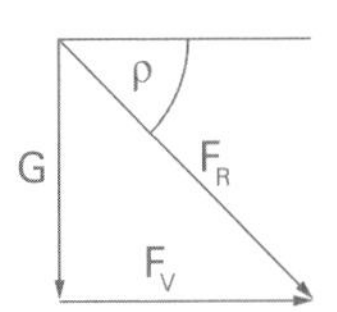

Robot with an axis length starting from axis 2 ca. 740mm, angle velocity 250° per second

Forces of inertia through rotation

If a workpiece is moved on a circular path, a constant acceleration towards the center of the rotary movement is necessary. This centrifugal acceleration triggers the counteracting centrifugal force.

$$F_Z = m\omega^2 r$$

As the square of the angle velocity enters the equation, enormous centrifugal forces arise. This especially applies to modern handling robots with fast rotary axes.

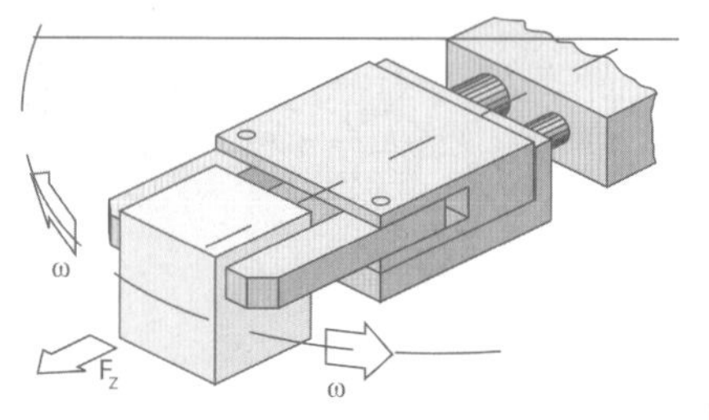

Figure 4.3 rotary force

As it is the case with translatory movements, the tangential acceleration at is necessary for changing rotary or angle velocity.
The tangential acceleration at acts perpendicular to the centrifugal force F_z or on the centrifugal acceleration a_z. The rotary velocity (ω) achieved in a set time t defines the so-called angle acceleration.

$$\dot{\omega} = \frac{(\omega_2 - \omega_1)}{t}$$

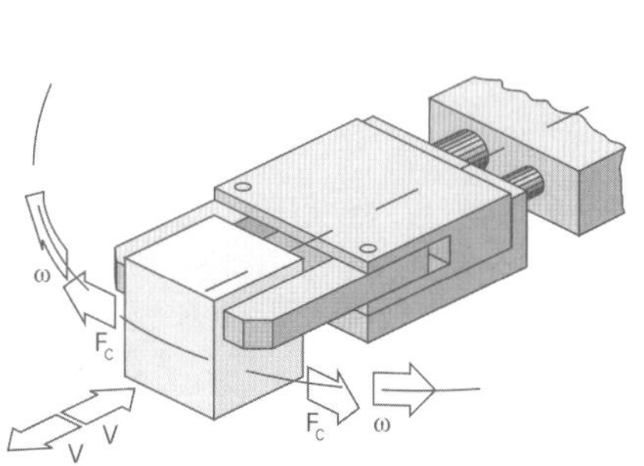

Figure 4.4 Coriolis force

An accelerated rotation will make the **centrifugal acceleration a_z** and the **tangential acceleration a_t** overlap.

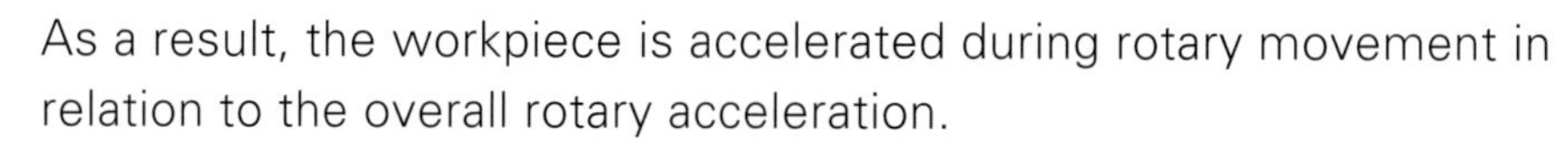

As a result, the workpiece is accelerated during rotary movement in relation to the overall rotary acceleration.

$$a_R = \sqrt{a_Z^2 + a_t^2}$$

with the direction angle

$$\beta = \arctan \frac{a_z}{a_t} = \arctan \frac{\omega^2}{\alpha}$$

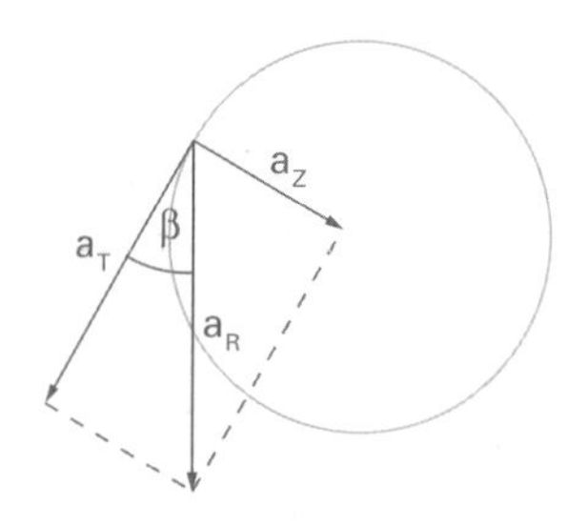

The Coriolis force

During rotary movements of a workpiece around the center point, the Coriolis force acts when the workpiece is moved towards this center point or vice versa. The Coriolis acceleration then acts perpendicular to the movement towards or away from the center point of rotation.

The Coriolis force is calculated as follows:

$$F_C = 2mv\omega$$

Influence of the gripping point

As explained in our analysis of workpiece variety (Chapter 3), it is ideal to grip the workpiece as near as possible to its center of gravity. This reduces the lever arm of the force generating the moment.

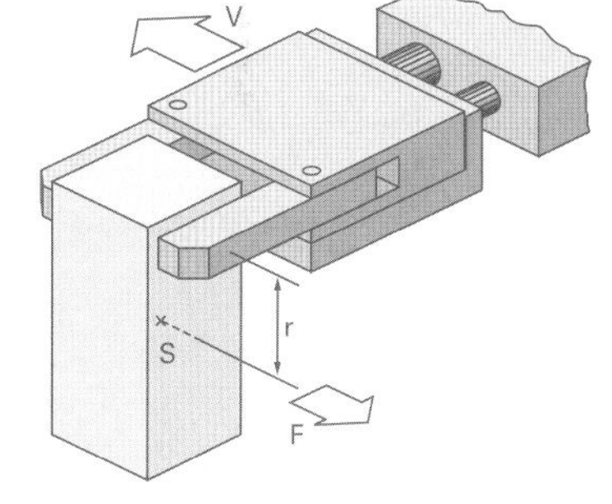

Figure 4.5 Displaced gripper fingers

$$M = mar$$

M = moment (Nm), m = handling mass (kg), a = acceleration (m/s^2), r = distance between gripping force and the workpiece`s center of gravity (m)

The moment counteracts the gripper`s operating elements which may lead to premature wear of gripper kinematics. In addition, the moment counteracts the movement of the gripper drive and, therefore, puts stress on the drive train.

As illustrated, the moment may lead to workpiece loss, in case the gripping force is calculated too low.

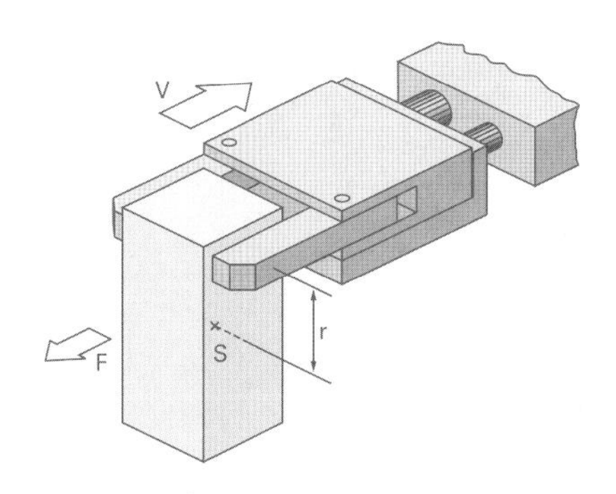

Figure 4.6 Slipping workpiece

Effects Of Forces Of Process

The forces of process are dependent on the respective handling tasks, such as workpiece assembly or processing. This type of force may also occur during pick operations, in case of sticky surface or bracing. These very specific forces must be calculated individually for each application and compensated as necessary. The options to compensate or even avoid this type of force are explained in the following.

If not only the position or orientation of a workpiece needs to be changed during handling, but also its quality, this is usually done by assembly or further processing steps. The forces acting on the workpiece are not necessarily forces of inertia because the movements are frequently too slow. When workpieces are polished, the forces acting on the workpiece and the gripper are dependent on the lever arm of the force as well as on the force required for pressing the workpiece onto the polishing disk.

The forces acting during workpiece processing must be calculated one by one in order to analyze their effects on the gripping force. With the help of modern technology, the forces of process can be measured. This option is used for many applications to increase system flexibility. If a force of process can be measured, a corresponding algorithm is usually found for actuating the kinetic device. The polishing process for differing blank workpieces, for example, can be regulated by the pressure of the respective workpiece onto the polishing disk. The FTC Force Torque Compliance sensor by ATI Theta Steifer KMS measures forces of process and ensures constant pressure of the workpiece onto the polishing disk. As a result, large-scale programming for complex workpiece shapes can be reduced.

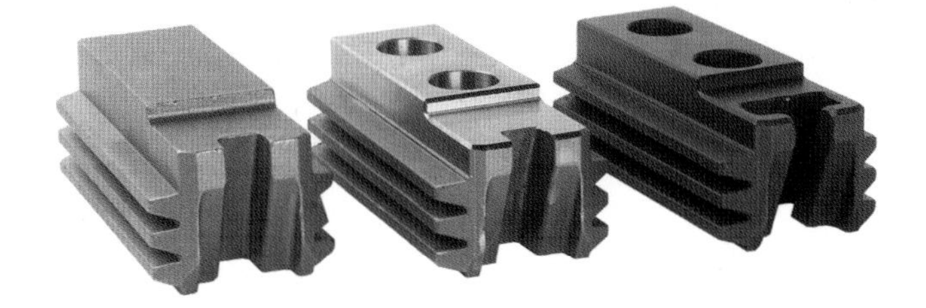

SCHUNK gripper guidances, processing steps (raw / processed / coated)

The quality of automated polishing has been greatly improved. This even applies to workpieces which formerly could not be polished due to their tolerances, such as castings.

Force of process during workpiece polishing (source: ATI Theta Steifer KMS)

Another example is controlling forces of process online during metal-cutting processing. As illustrated, the robot can vary the pressure with the help of sensors fitted to the drill bit, which measure the cutting forces in relation to drill diameter and tool wear.

These examples are an indicator for today`s potential in sensor technology, measuring forces of process. The latter are most interesting in terms of automated solutions for workpiece assembly.

Forces of process temporarily occuring during emergency stops, when movement comes to a sudden halt, are a matter apart. The delays, which can be achieved, are frequently higher than the acceleration of the axes and, therefore, are most important in terms of operational safety.

Robots with fixture for drilling (source: ATI Theta Steif, KMS)

An emergency stop may occur at any moment of a kinetic situation, which makes construction difficult. It is recommended to conduct a risk analysis for the respective kinetic situation. Appropriate safety provisions are necessary according to the workpiece`s weight and form, such as form-fit grippers, protective screens, or kinetic changes.

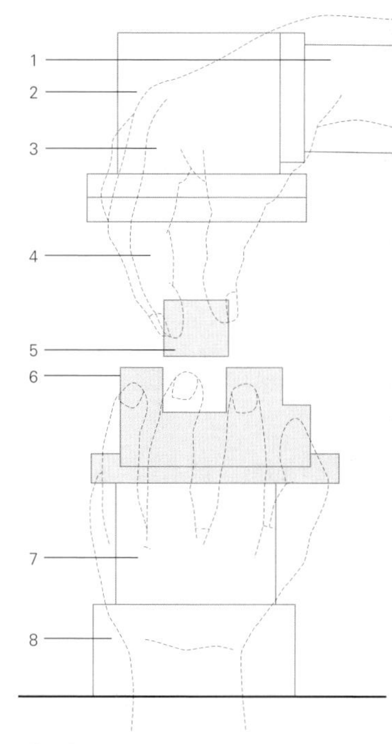

1 *robot arm*
2 *assembly mechanism*
3 *gripper change system*
4 *gripper jaws*
5 *workpiece*
6 *workpiece base*
7 *rotary unit*
8 *sensor unit*

Figure 4.7 Manual assembly of workpieces analog to the technical process

Assembly tasks still require a great deal of manual work because workpiece variety is vast while lot numbers of identical workpieces tend to be small. As a result, assembly requires highly flexible technology. A simple task, such as putting a bolt into a hole, illustrates the problems of automated assembly. Tolerances in workpiece positioning in relation to the workpiece base, make the assembly difficult if not impossible at all. Assembling workpieces without rotary symmetry is even more complicated.

If the position of a workpiece is not adjustable to the position of the workpiece base, they cannot be assembled. Humans with tactile and visual perception start learning to perfectly solve this kind of task in their early childhood. Sensor systems and passive compensation systems, which actively compensate faulty positioning, have been developed for such challenging tasks.

Reducing And Measuring Process Forces

Passive compensation systems generate the compensation of positioning errors from the movement of the kinetic system alone.

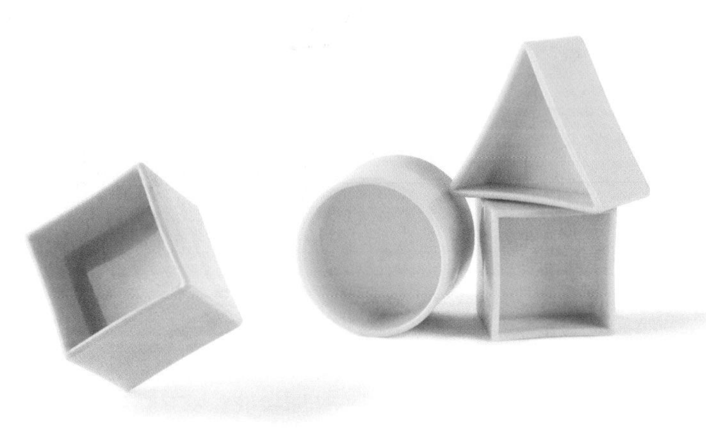

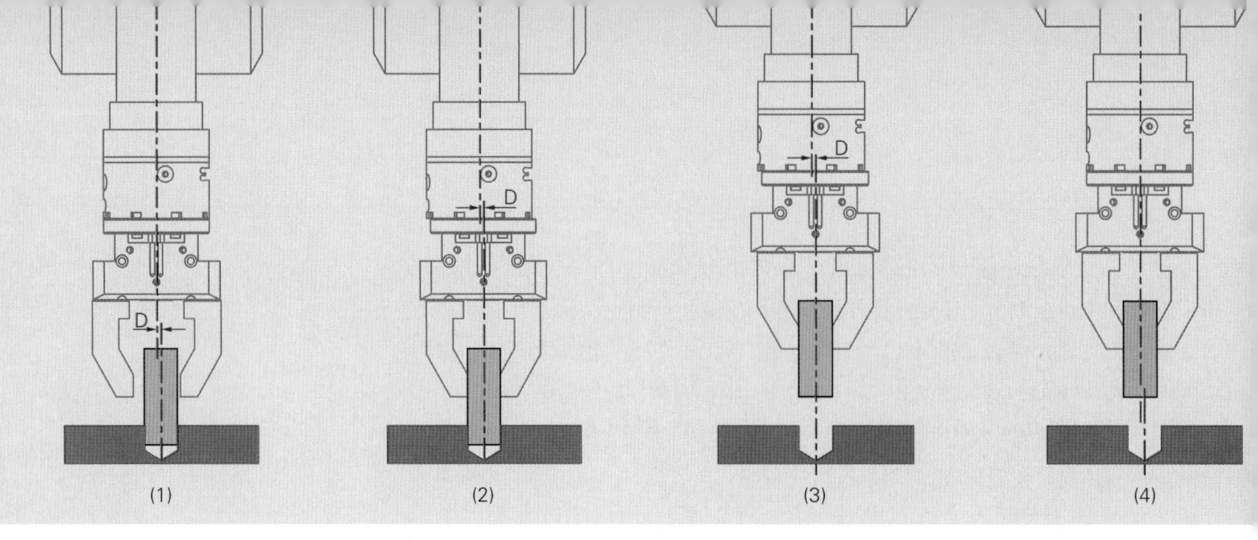

Figure 4.8 Principle of function: Compensation unit removing a bolt from its faulty position

AGE compensation unit

The **AGE** compensation unit enables passive compensation of position errors along the x- and y-axes up to ±4mm depending on the size of the component. Faulty angle positioning can be compensated up to 16°. The compensation unit can be pneumatically locked to keep it stable during robot movement errors. It is possible to lock the center position or any other position. Faulty positions caused by robot teaching, for example, can be compensated and "saved". This reduces the force which acts on the robot and gripper during pick operations.

Magnetic sensors can be fitted to tongues to make sure that the compensation unit has been locked.

If the compensation unit is directly fitted to a robot flange it is easier to integrate the component. ISO 9409 specifies drilling for flanging components which makes the flange adapter plate redundant. This reduces weight and cost.

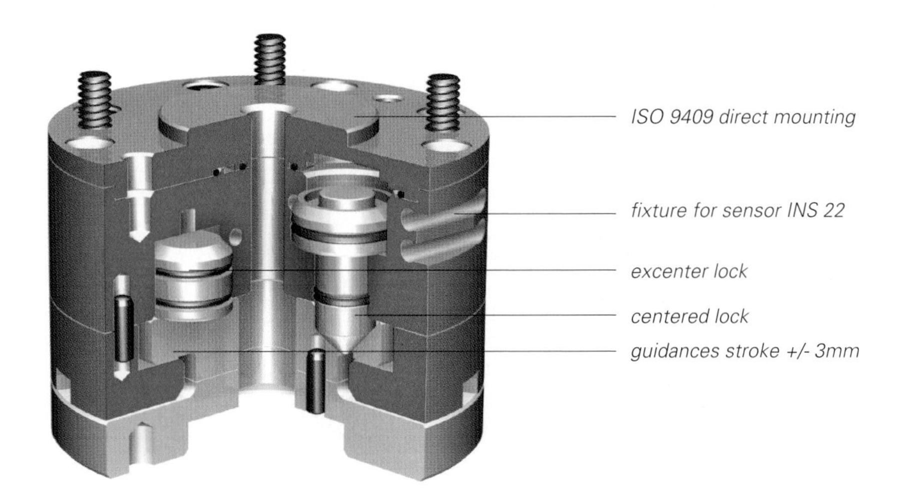

Active compensation of faulty positioning during assembly or reaction on forces of process is realized by **FT** (**F**orce **T**orque) sensors. These sensors optically measure forces and torques and convert them into signals for robot drives

Six diodes are mounted to the core of the sensor and send their light through a cover plate ring to the **PSD** (**p**osition **s**ensitive **d**etector) which record the change in position. These data are converted into kinetic data, which are the basis for calculating the forces and torques if the counterforces of the integrated spring packs are known. The latter are visible in the sectional view of the sensor. An additional option to lock the sensor must be provided, in case the robot must perform a fast pick-up movement. This can be achieved pneumatically for the sensor depicted.

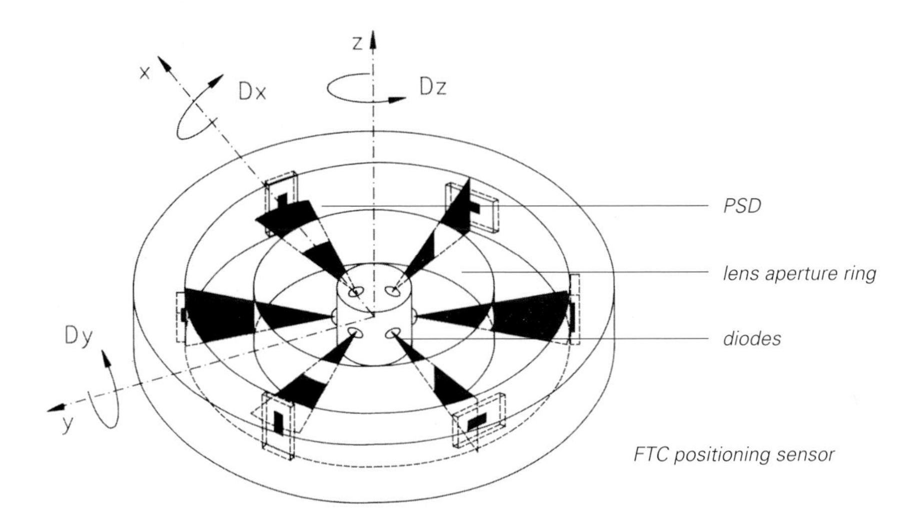

FTC positioning sensor

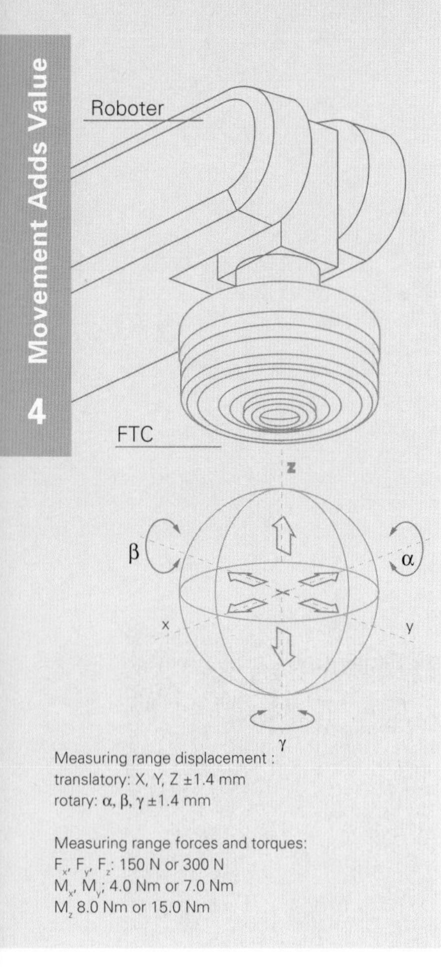

Figure 4.9 FTC principle of function

FT sensor in operation

This sensor can measure forces up to 300N and torques up to 15Nm taking into consideration the directions of force and torque. Workpiece displacements which the sensor tolerates are maximum ±1.4mm for the directions x, y, z, and maximum ±1.4° for the rotary directions α, b, γ.

The data measured by the sensor can be exported by CAN, DeviceNET as well as by RS232 or RS485. The data are updated per millisecond in each case.

The sensor comes with a PC compatible test software for checking all functions and putting it into operation. All sensor functions can be triggered by a simple parameter input. No extra drivers are required and the test software is compatible with any software.

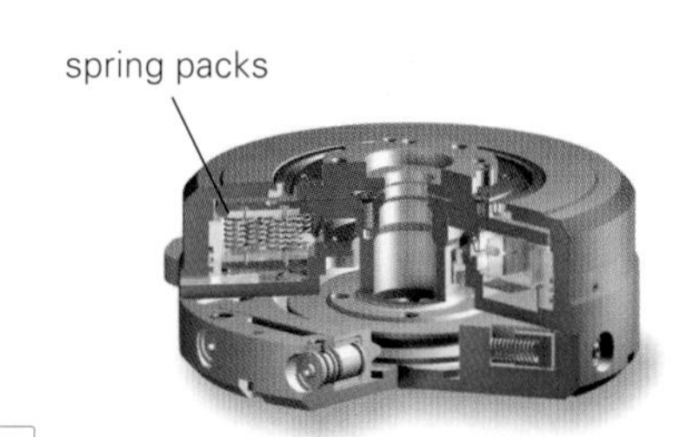

FTC sensor, sectional view

Another type of FT (Force Torque) sensor is based on a tactile measuring procedure. As illustrated, the forces and torques are measured with so-called DMS. The tool is fitted to the inner ring of the sensor. The tool (or gripper) force is transmitted by three crossbars onto the fixture ring of the robot.

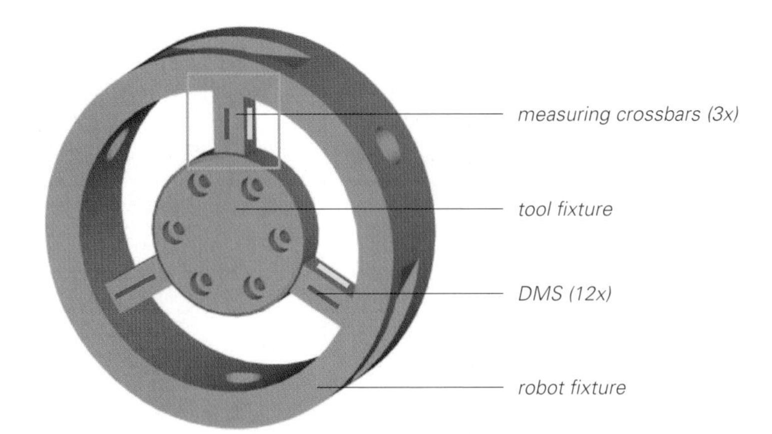

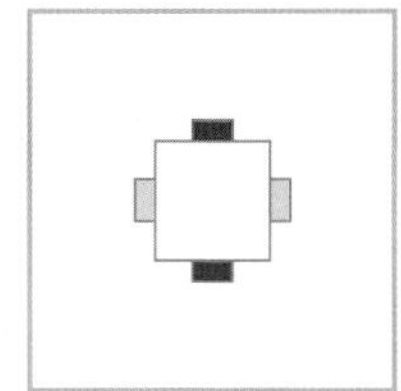

measuring crossbar, sectional view with DMS strain gauge sensors

Manifold applications have been realized with this principle of measuring, ranging from mounting sport coupé roofs to assembling kitchen sinks. Monitoring and controlling forces of process makes these applications safe or apt for automation in the first place.

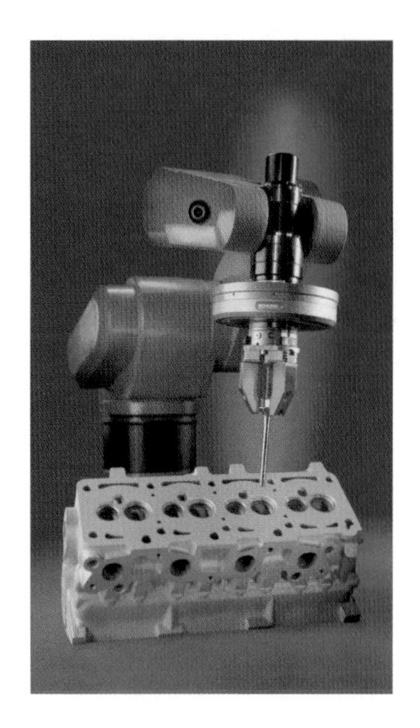

sensor used for working on a cylinder block

Exceptional Forces Of Process

A collision with an obstacle in the workspace is considered a special case of forces acting on the workpiece/gripper combination. Such collisions tend to occur during implementation or for particular robot movement patterns which are difficult to test. For example, sensor signals may change the kinetic path of a robot if the safety provisions not sufficient.

Collisions are frequently provoked by human operator failure in case obstacles, such as pallets or the wrong workpiece types, are placed into the robot workspace. In most cases, so-called anti-collision units immediately pay off, considering the costs for loss of production without even counting machine damage.

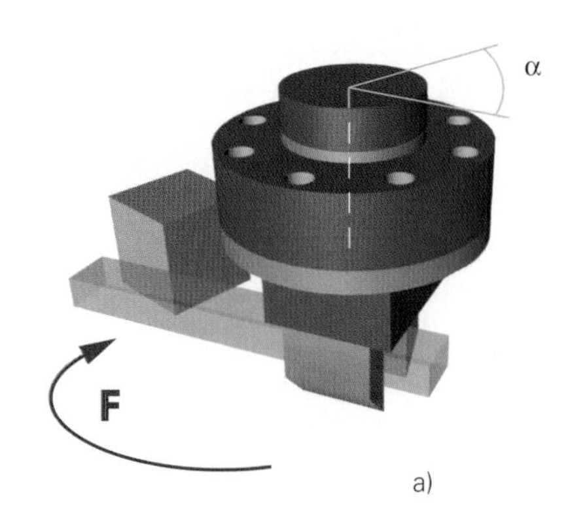

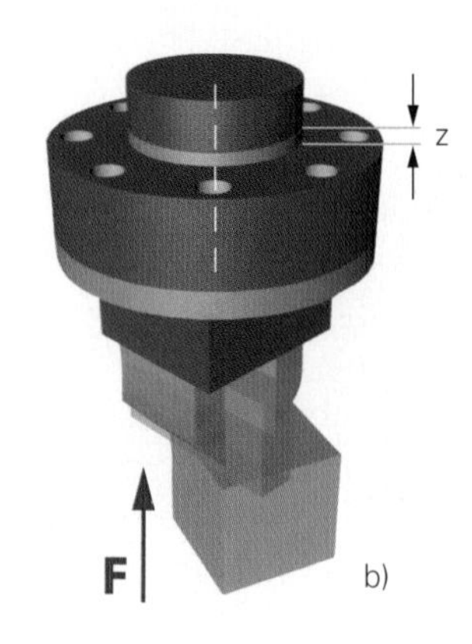

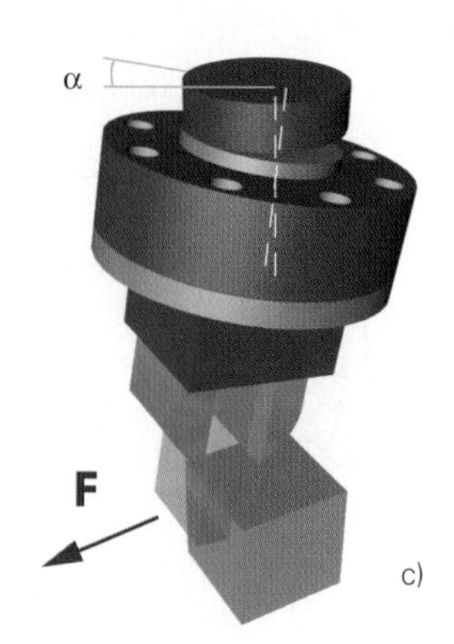

Figure 4.10 Examples of collisions with the respective directions of force and torque

Such anti-collision units must trigger an emergency stop if forces exceed the limits of normal operation. Figure 4.10 includes different examples of possible collisions and related forces or torques acting on the gripper and the protection unit.

In example a) torque acts on the gripper and thus on the anti-collision unit as if the gripper is to be rotated. Example b) shows pressure acting on the gripper from below which is typical for assembly. Example c) explains a very common situation where torque may cause the gripper to tilt.

The anti-collision unit is kept rigid by air pressure during normal operation. By changing the input air pressure by a throttle, various degrees of rigidity can be adjusted in relation to the workload.
As soon as overload or collision is detected, sensors integrated into the overload protection give a signal, which then triggers the emergency stop at the respective process control. For regular protection, the piston is switched without pressure or the air in the piston space is released by a throttle check valve with reset spring. The drawback of the latter is that an abrupt release of air is not acceptable for certain applications, such as cleanroom.

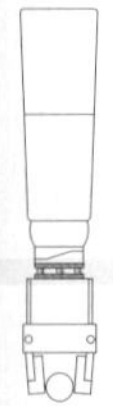

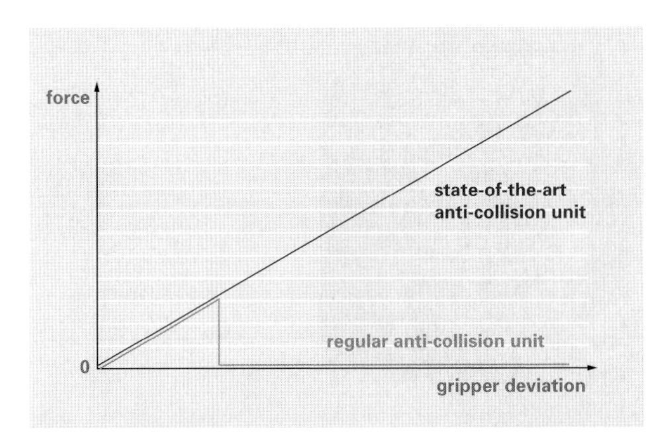

Figure 4.11 Force-deviation diagram for anti-collision units

Modern anti-collision units generate the maintenance force through spring packs with the result that the force can be kept linear with gripper deviation over a long distance. Two different types of anti-collision units with their respective force-distance curves are illustrated in figure 4.11.

The curves represent force over gripper deviation with the red curve showing the linear ascent of force over deviation. The latter is possible, if the anti-collision unit is solely kept in position by spring force. The curve of the regular anti-collision unit drops to zero after having reached its force maximum, as it is typical for anti-collision units supported by a pneumatic cylinder.

Modern anti-collision units do have several adjustment options. Emergency stop signals can be individually set so that, for example, the anti-collision unit tolerates a deviation up to 3mm before it gives the signal. This permits the operator to adjust the emergency stop function according to the desired timing or degree of deviation. This is of advantage for applications which permit slight gripper deviations but risk damage in case of great deviations.

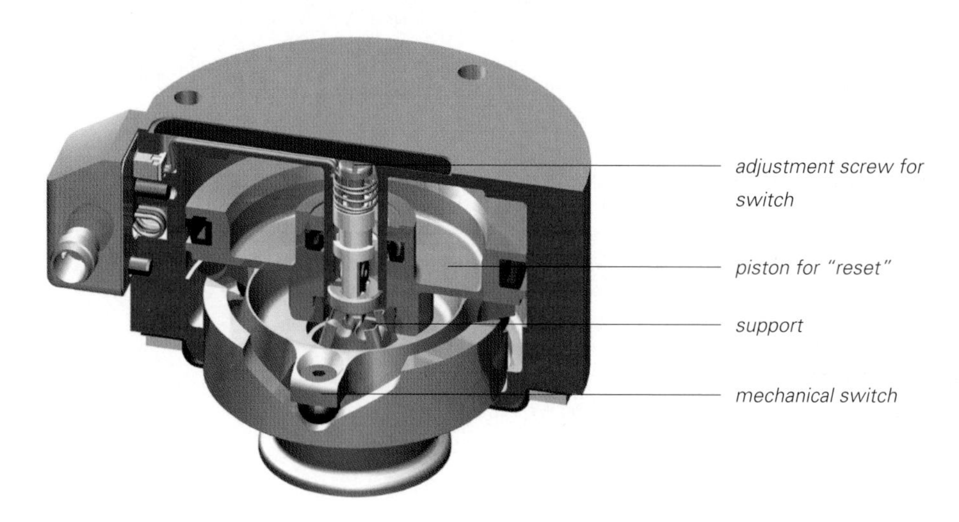

anti-collision unit, sectional view

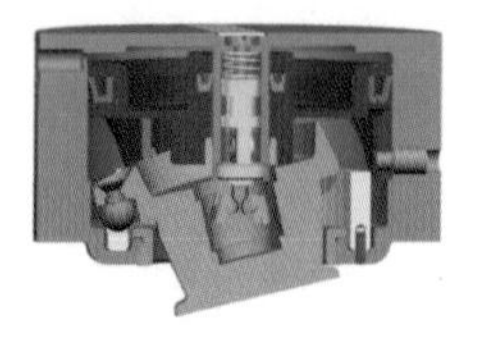

maximum deviation

Anti-collision units can also be automatically reactivated after collision. The cylinder can center the gripper again after renewal of pressure, which is interesting for applications where a collision does not necessarily require an emergency stop. As a result, a pick- or place station can be started again without the help of an operator and there is no need to enter the workspace of the kinetic unit.

This type of anti-collision unit can be used as a compensation unit as well. It is equipped with a solid support, which hardly wears even at frequent use.

So-called overload protection units, which are again based on spring force, are used for some applications. These units show different force-deviation diagrams in relation to the direction of stress, as illustrated for a specially compact unit.

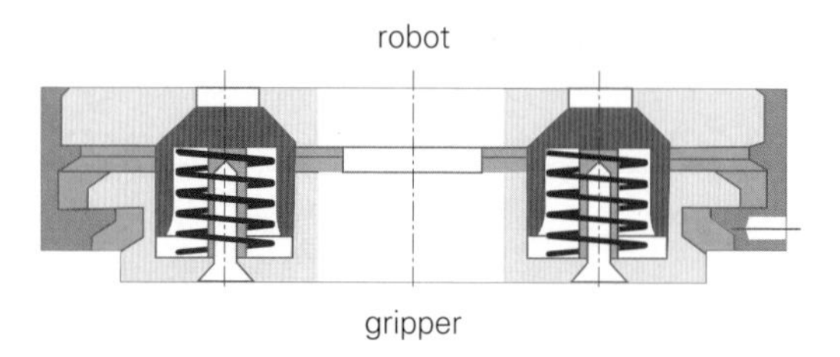

overload protection unit based on spring force

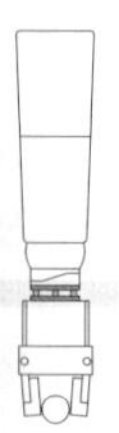

4.2 Realization of Kinetic Processes

Workpiece movement is expressed by Newton`s kinetic equation. A movement in space is basically defined by its velocity, acceleration, and direction. In automation technology, movements are realized in different ways. All basic options are listed in a structured overview in the VDI Guideline 2860:

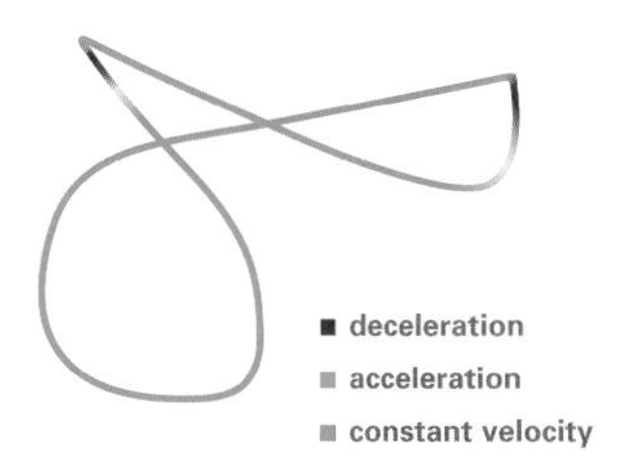

Kinetic path with acceleration and deceleration in different colors

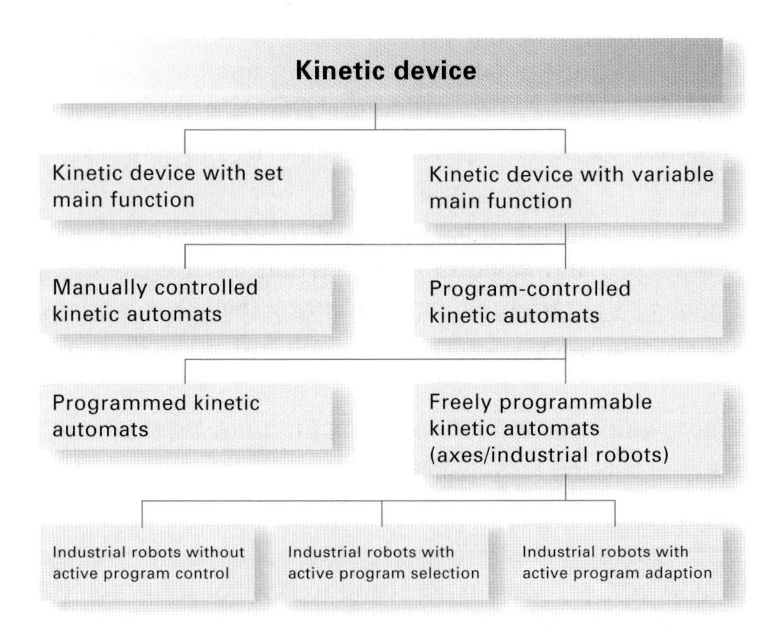

Figure 4.12 Kinetic device options (VDI 2860)

Generally speaking, a kinetic device with a set main function is a low-cost option for changing workpiece position or orientation, such as pneumatic cylinders and mini slides for workpiece positioning. Kinetic devices with variable main function include both manually controlled and program-controlled kinetic automats.

Variable main function means that movements, which are performed by the kinetic device, are not pre-defined but can easily be changed. Manually controlled automats are not explained in detail in this book. In short, they are usually manipulators used for movements which would be too health-damaging for humans to perform, such as handling heavy weights in palletizing or handling radioactive elements in nuclear power plants.

Program-controlled kinetic automats and their sub-systems are detailed in this chapter. Programmed kinetic automats, such as curve-disk controlled kinetic automats, have been taken into account as well as freely programmable kinetic automats, which are represented by PC- and NC-controlled axis movements.

The freely programmable kinetic automats are subdivided according to their degree of "intelligence". Increasingly independent kinetic devices change their programs in terms of behavior and adaption to changing environments.

They either change them by selecting a program from a number of defined sub-programs or by adaption, i. e. changing their program or their movement patterns.

Program adaption is widely used for robots which need to organize their movements in unstructured environments without special foreknowledge.

Simple Rotation
First we take a look at kinetic devices with a set main function, i. e. components performing the most simple translatory or rotary movement in order to reach an end position.

Swivel unit

The so-called **rotary or swivel units** with their rotary movements are used

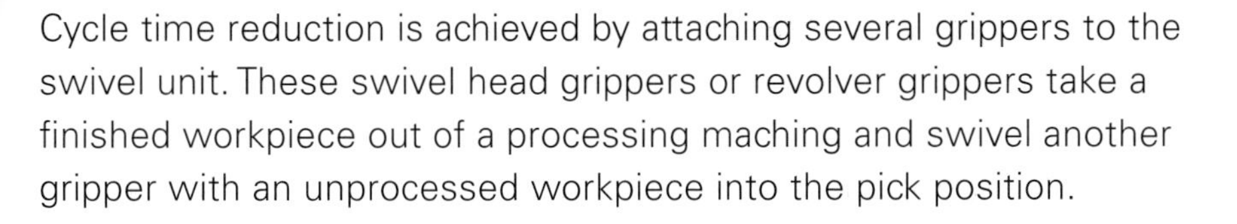

1. as components for multiple grippers to reduce cycle time and
2. as a kinetic device for workpiece orientation

Cycle time reduction is achieved by attaching several grippers to the swivel unit. These swivel head grippers or revolver grippers take a finished workpiece out of a processing maching and swivel another gripper with an unprocessed workpiece into the pick position.

Auxiliary process time for feeding the processing machine is reduced to a minimum as the swivel movement can be performed without any or with minimum effort by the handling device. By using the swivel unit for pick operations, the handling device is able to avoid movements that would pick up an unprocessed workpiece.

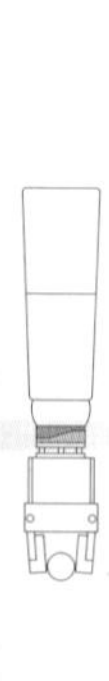

SCHUNK

Figure 4.13 illustrates that the processing machine has to wait for the handling system to place the workpiece onto the pallet for finished workpieces, if it cannot fall back on a swivel unit. Thereafter, the handling system must pick up an unprocessed workpiece and take it to the processing station. Handling time for workpiece pick- and place operations must be added to processing time, as all operations take place in sequence. Handling systems with swivel unit operate parallel, i. e. synchronized to the actual processing. Thus processing machine capacity can be increased in relation to set-up and velocity of the handling system by synchronization. The objective is to have the handling system with the unprocessed workpiece ready before the processing machine has gone through one cycle.

The swivel unit can be utilized for workpiece orientation as well. This option is used when a handling system does not offer sufficient degrees of freedom, for example, if a four-axis robot requires the workpiece to swivel into a horizontal position.

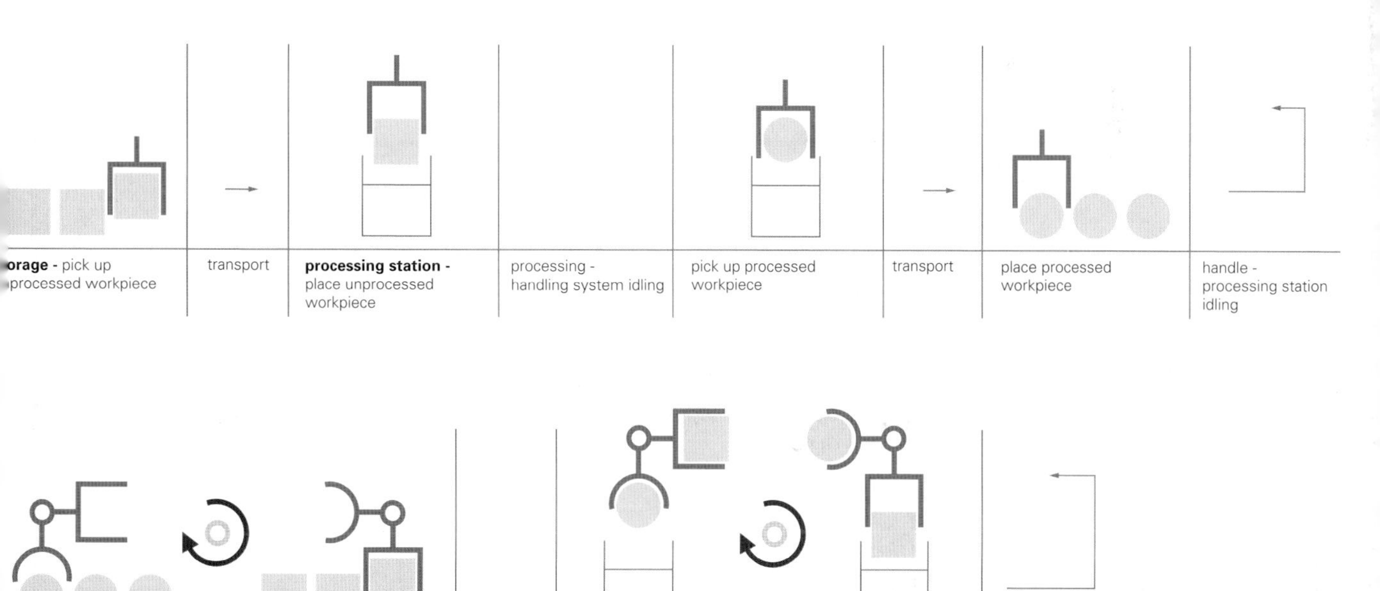

Figure 4.13 Feeding a processing machine without swivel unit and with swivel unit

direction of rotation

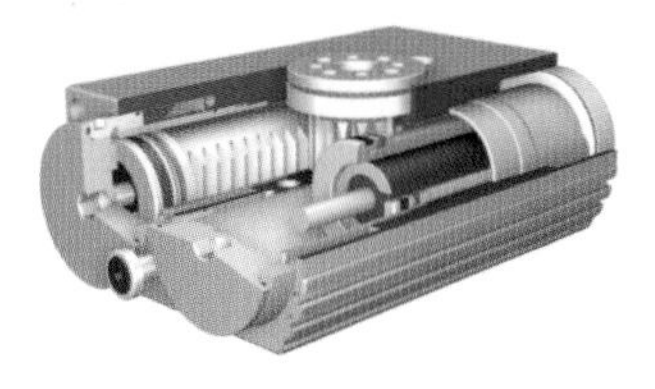

Swivel unit, CAD sectional view

Some applications require flexible rotation angles. In order to meet this demand, swivel units have been developed, which are freely programmable.

These swivel units permit a variety of swivel angles and directions for setting up another gripper to handle the workpieces. With swivel units based on the principle of pinion tooth rack, various positions can be selected according to the required rotation angle.

Figure 4.14 Swivel unit based on principle of pinion tooth rack

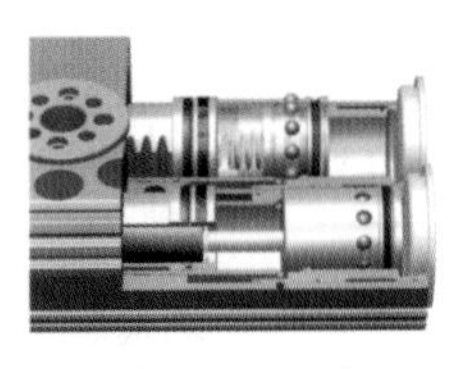

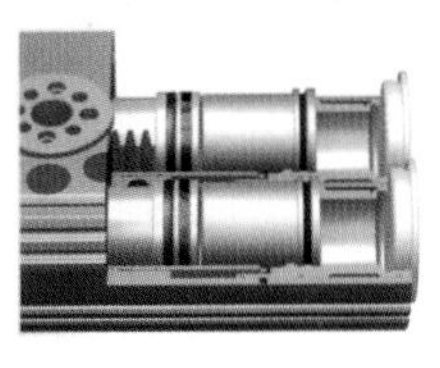

Swivel unit with (above) and without (below) mid position

An additional piston will make mid positions possible if required. This swivel unit option is fitted to the basic swivel unit.

Three options for rotation angles are available:

- swivel units with two positions and fixed rotation angle, which the user can pre-select (fig. 4.14)
- swivel units with two positions and flexible rotation angle, which the user can freely adjust (fig. 4.15)
- swivel unit with two positions and mid position, which the user can freely program (fig. 4.16)

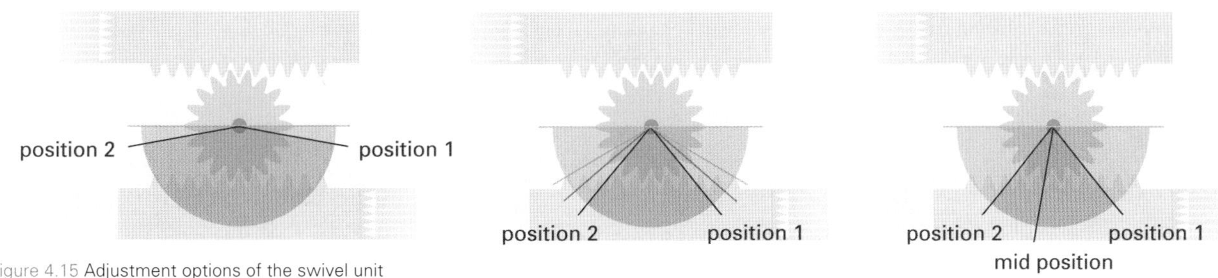

Figure 4.15 Adjustment options of the swivel unit

The rotation angles are generally limited to 90° or 180°, which means that the rotation of the swivel unit does not go beyond these maximum positions due to its type of construction. Swivel movements in a closer range can be adjusted by altering these end positions.

For the mid position the same kind of adjustment is possible +/-3°. The mid position can also be locked to make it a safe stationary position.

The end positions are hydraulically dampened because a pneumatic solution, especially for short swivel times and great mass moments of inertia, would lead to hard impacts and premature wear.

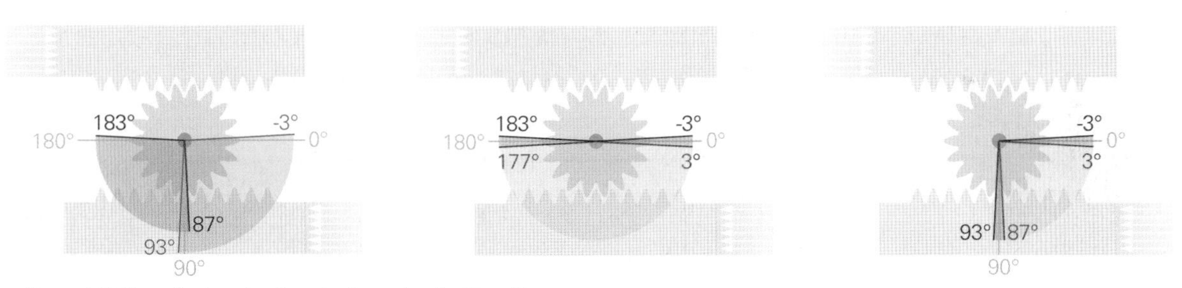

Figure 4.16 Fine adjustment options for the end and mid positions

It is possible to make fine adjustments to the end positions of the swivel units in order to precisely position the swivelled gripper.
If the end positions are infinitely variable and accurately set, better adaption to gripping situations is achieved.

Magnetic switches can be used for calling up to eight positions of the swivel unit. The electronic magnetic switch can be completely integrated into the housing so that interference contours hardly occur. Initiators can be used as an alternative; they are larger in size but offer higher operational safety.

Initiators can be used as an alternative; they are larger in size but offer higher operational safety. Inductive proximity switches have three positions, which are triggered by a cam at the swivel unit.

When employing swivel or rotary units, the following parameters have to be taken into account:

- torque
- mass moment of inertia
- swivel time

These parameters determine the rotational enery saved in a rotary movement.

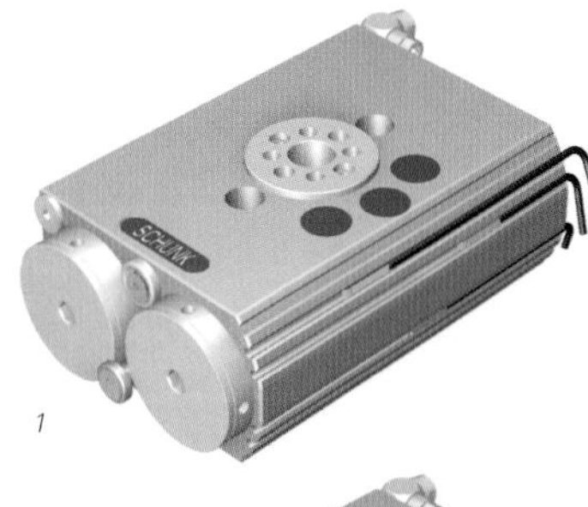
1

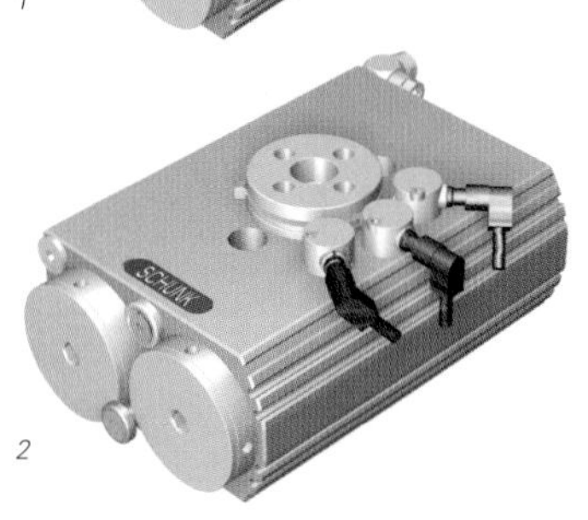
2

Swivel unit with electronic magnetic switches (1) or initiators (2)

Rotational energy

$$E = \frac{1}{2} J\omega^2$$

$$\omega = \frac{\lambda_1 - \lambda_2}{\Delta t}$$

E = rotational energy (Nm), J = mass moment of inertia of the swivel head (kgm^2), ω = angle velocity (1/s), λ = rotary angle (rad), t = swivel time (s)

The rotational energy defines the maximum torque, as stated in the product specifications in relation to the size of the swivel units.

The larger a swivel unit is in size, the larger the maximum torque becomes. As illustrated, especially high torque is achieved through double actuation of the cylinders. For the single actuation, only one piston is used for generating the swivel movement. Double actuation achieves a torque more than twice as high for the same swivel unit size.

Calculations for the necessary torque are different for rotary units and swivel heads. The respective mass alignment towards the rotary center point change the torques required.

The depicted swivel units achieve a specially high torque by the double actuation of the cylinders.

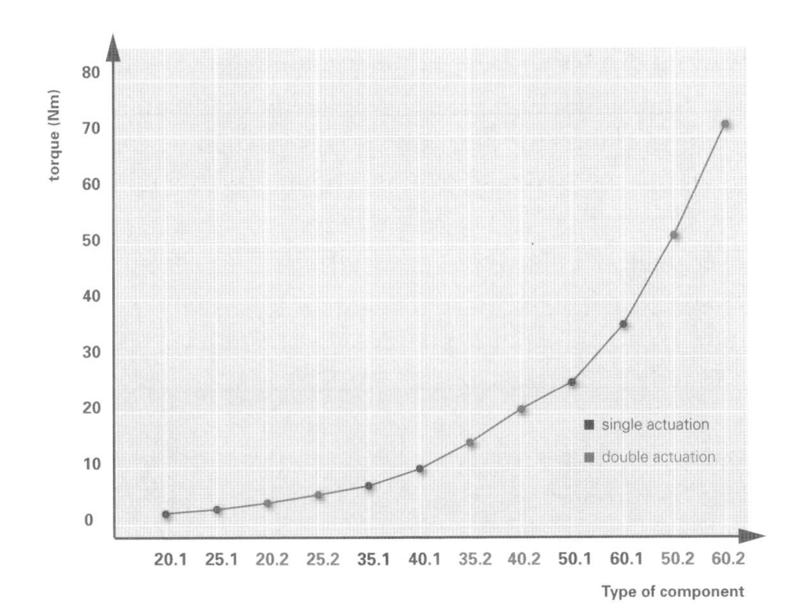

Figure 4.17 Graduation of torque for different swivel unit sizes

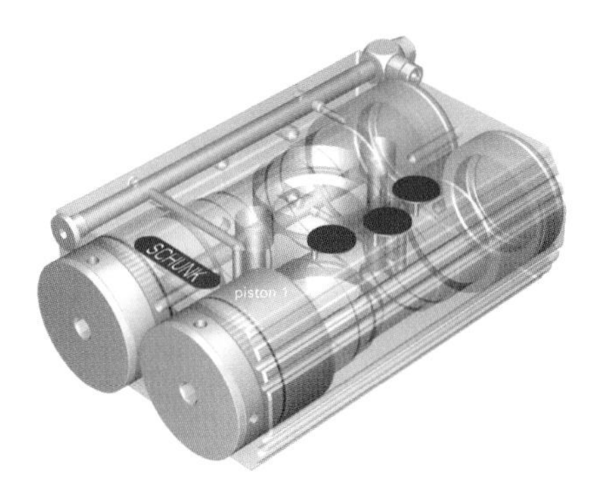

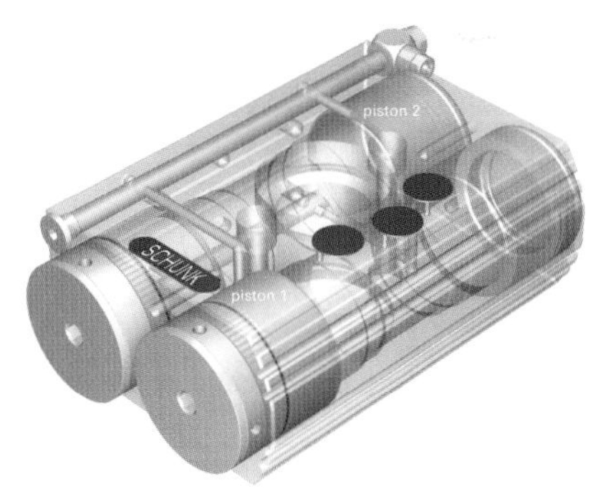

Air feed-through at single actuation. One piston actuated with air. (left)

Air feed-through at double actuation. Both pistons actuated with air. (right)

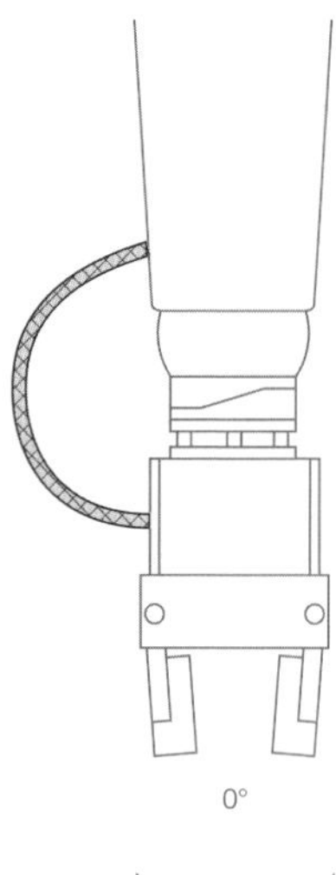

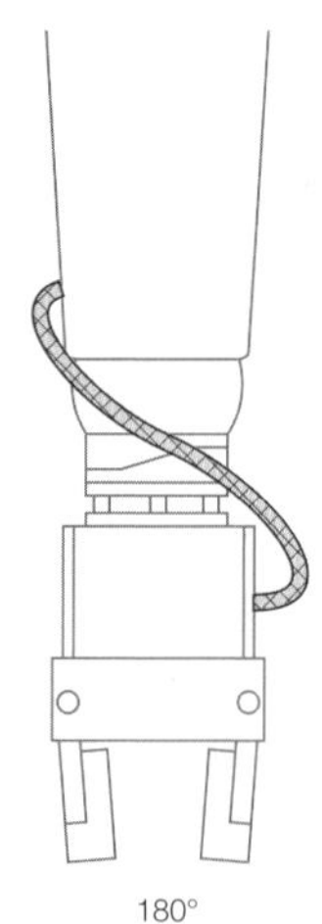

Figure 4.18 Rotation of the gripper, external guidance of tubes

The swivel or rotary time depends on the acceleration of the mass and on the translation ratios, i. e. the size of the swivel units. Product specifications generally state the swivel times achieved without load. Swivel times required for a rotary movement usually vary from 0.04 seconds to 1.0 second.

Cables and Tubes – Obstacles of Movement

One major problem caused by rotating grippers and swivel units is obviously the energy supply. Cables and tubes, which are constantly subject to torsional stress, are bound to brake. This problem can only be solved by an integrated, tubeless energy supply.

Rotary distributors can either be electric or pneumatic. Electric energy can be transmitted from the mount-flange to the gripper. Air pressure can be transmitted through bushings, and up to eight bushings can be utilized depending on the type of swivel unit.

Not only swivel units but also robot axes face the problem of tubeless energy supply. Energy is indispensable for the gripper to be able to operate. It can be electric, pneumatic, or hydraulic, in accordance with the type of drive selected.

There is an increasing need to transfer information in the form of electric signals from the gripper back to the kinetic device.

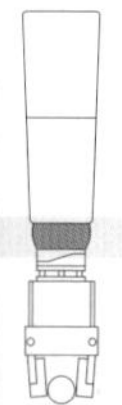

The torsion of the gripper in relation to the main axes occurs at the so-called hand axes of a robot. The hand axes are able to twist the gripper, i. e. change its orientation in relation to the basic kinematics. These torsional movements may cause not only the energy supply tubes to twist but also the cables leading to the sensors.

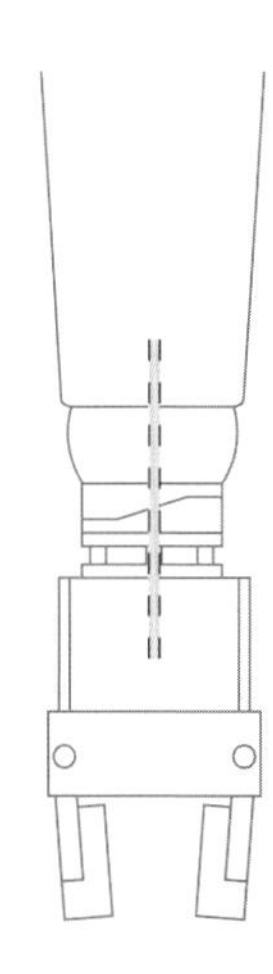

Figure 4.19 Rotating movement of a gripper with rotary distributor

Component producers offer rotary distributors to transmit the energy provided at the robot flange to the gripper without the use of cables or tubes. These rotary distributors can transmit the type of energy appropriate for their construction.

The rotary distributor illustrated here has four pneumatic bushings at maximum 10 bar. In addition to the tubeless energy transmission, up to ten electric signals at maximum 60V and 1A can be transmitted by an electric bushing via a slip ring. As a result, rotary movements are not disturbed by any cables or tubes.

CAD sectional view: rotary distributors

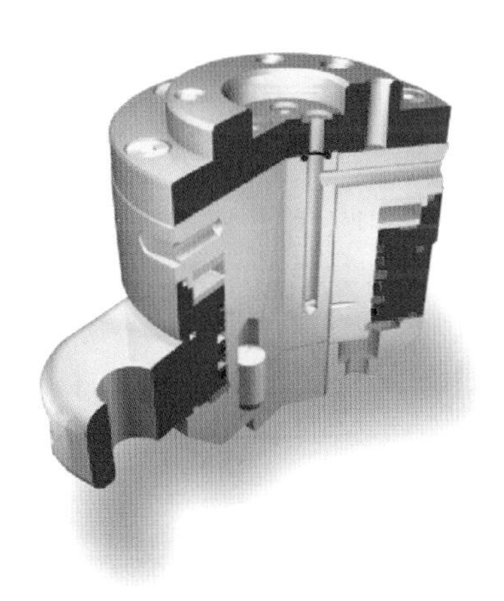

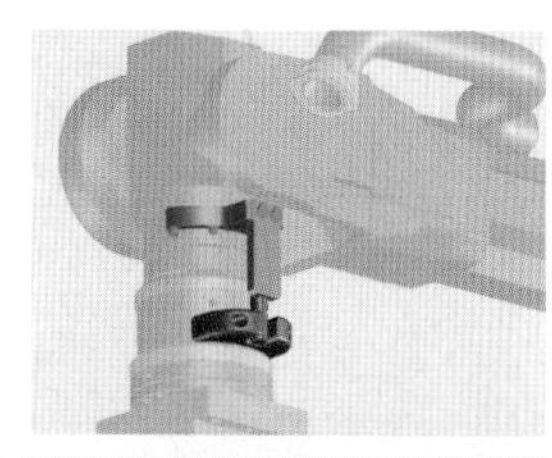

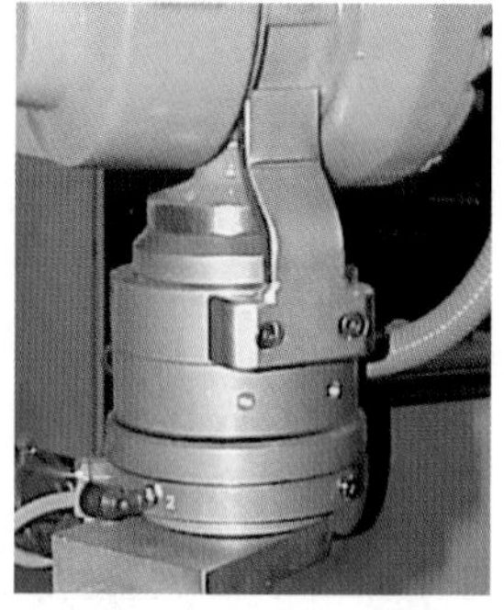

Rotary distributor with torque reactor strut

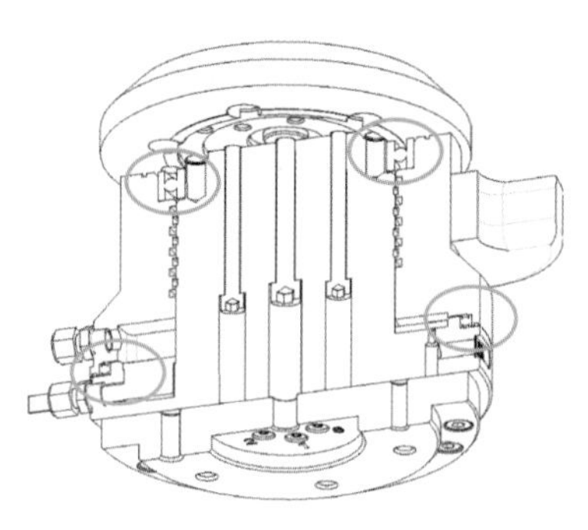

Rotary distributor with ball-bearing (marked in blue)

Rotary distributors for fluidic or electric energy have become frequent components of grippers over the past few years. System errors were formerly caused by faulty or leaking air pressure tubes or broken electrical wires. This kind of problem occurred with swivel units, rotary units, and robots, and was solved by standardized rotary distributors for most diverse tasks. For special requirements, such as very high currents or manifold cabling, customized solutions are available.

All rotary distributors have one or more inputs which do not perform a relative movement in relation to the rotary movement of the robot or the swivel unit. The so-called torque reactor strut against the main axes of the robot kinematics is illustrated below.

The reactor strut prevents the input of the rotary distributor to move with robot movement, which may be caused by internal friction in the rotary distributor. In order to reduce this internal friction to a minimum and to transmit great forces at the same time, the rotary distributors are available with mounted ball-bearing.

The housing plays an important role in determining which types and quantities of energy can be transmitted. One example is the rotary distributor at 1,500A, which is required for robots welding component parts.

Special rotary distributors with 6 pneumatic and 12 electric bushings with gripper change system and torque reactor strut

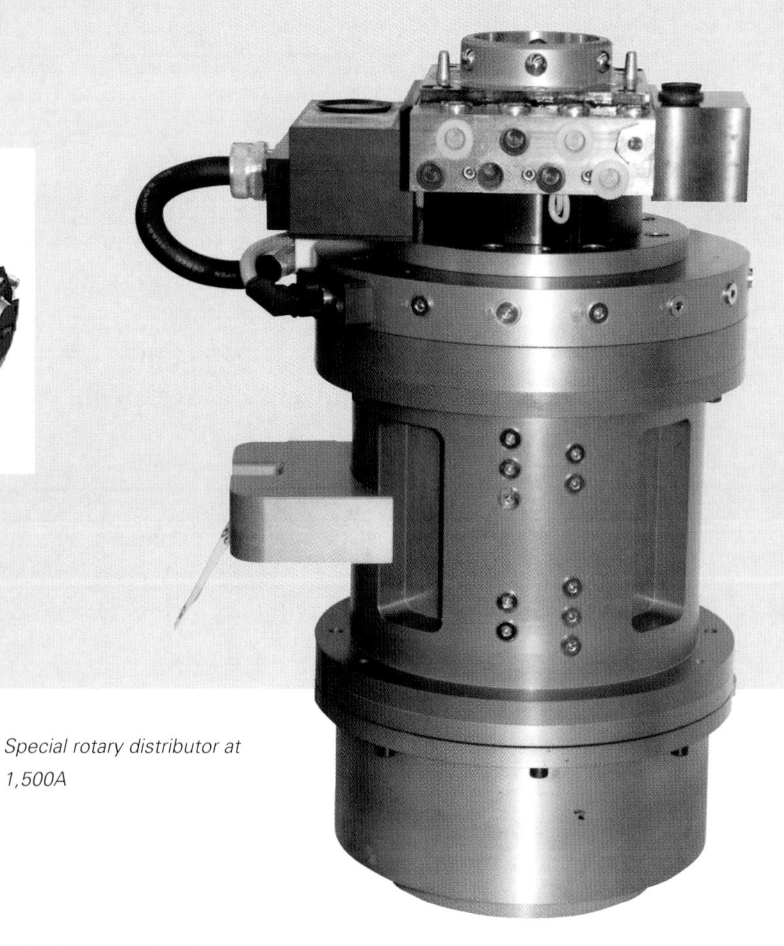

Special rotary distributor at 1,500A

Pneumatic transmission is problematic for applications which do not permit valve technology to be integrated into the gripper. As a result, all switch operations must be performed within the reach of the robot arm, or even further away and separately through the rotary distributor to the gripper.

Transmitting information from the gripper to the robot becomes more and more important with the increasing number of sensors. Rotary distributors make this transmission possible without twisting any cables.

An alternative for transmitting information from the gripper to the control of the kinetic device is wireless transmission technology. The latter can be expected to greatly reduce the amount of cabling in automation over the next few years.

Change of Grippers

If the energy supply must be completely separated because the whole gripping unit needs to be changed, so-called gripper change systems are employed. They are mounted to the flange plate of the robot and gripper system and can be easily separated, both mechanically or in terms of energy supply. The gripper change systems are adapted to the interfaces defined according to DIN ISO 9404.

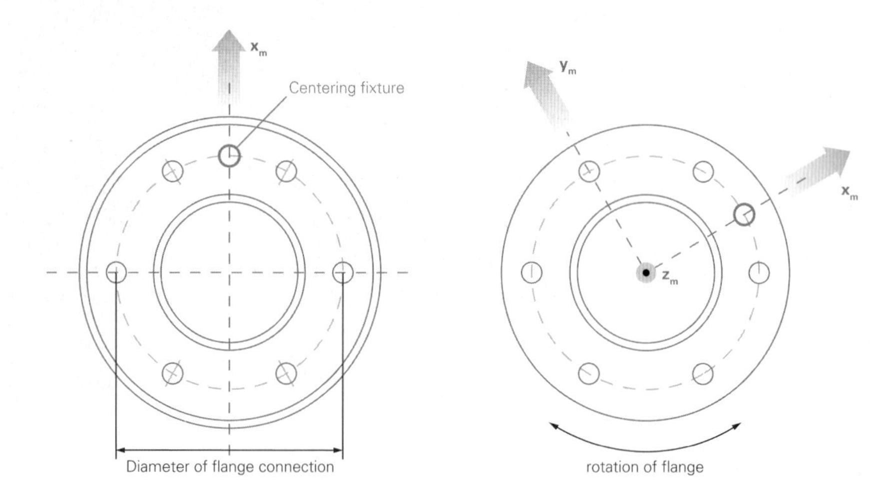

Figure 4.20 Flange plate according to DIN ISO 9409 and flange system of coordinates

The task depicted seems quite plain at first sight but turns out to be a major challenge when looking at the details. Not only the gripper must be mechanically separated from the robot flange, but also the energy supply and the information channels need to be separated. This procedure must be automatic and without faulty contacts or leakages, which means that high precision is expected from the connection.

The electric energy is usually transmitted in a module specially designed for this purpose. The mechanical lock for this gripper change unit is realized by a pneumatic drive.

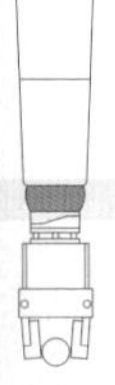

SCHUNK
SCHUNK
SCHUNK
SCHUNK

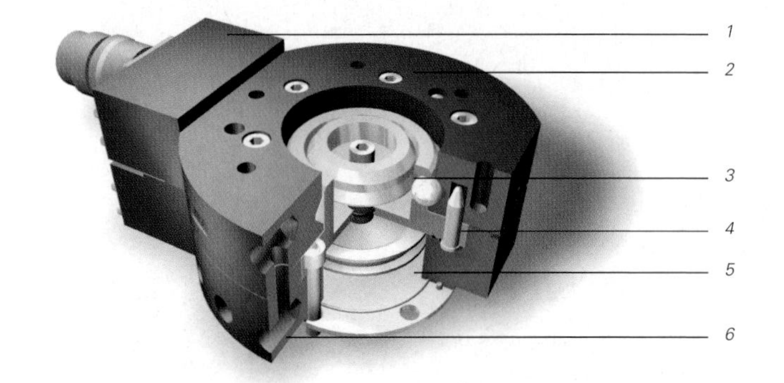

1 Electric module
for electric transmission of energy

2 Housing
optimized weight by using a high tensile aluminum alloy

3 Mechanical lock
force-free unlocking and locking with self-locking feature

4 Pneumatic rotary distributor
no interference contour by integration into the housing, adaptable for vacuum

5 Positioning pins
for accurate operation of the clutch and high precision

6 Drive
pneumatic for efficient and simple operation

The option of separating the gripper from the kinetic device increases flexibility by far (see Chapter 3). Thus the kinetic device can be used for different tasks or workpieces and can perform more than just one main function. One of the pre-conditions is that the kinetic device is able to move to more positions than just one for taking up the new gripper. The latter is hardly possible for most kinetic devices with one defined main function.

Gripper change unit

Linear Movements

Kinetic devices with a defined main function include swivel units as well as simple components for performing linear movements. The gripper is moved between the end positions of the kinetic device. The so-called mini slide is a good example of a component for this kind of task.

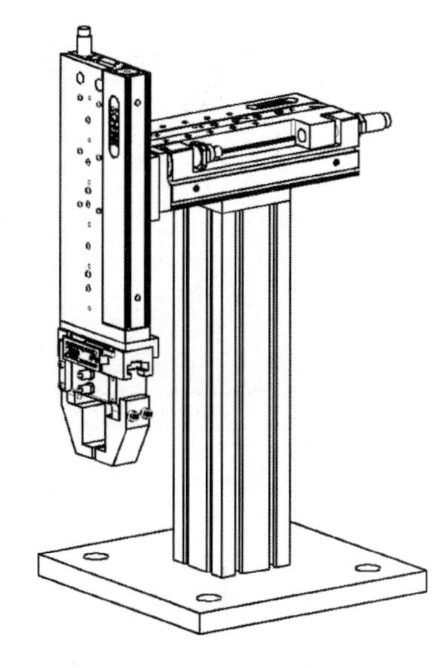

Mini slide system

The mini slide stands out for its good guiding properties of the translatory movement. High forces can be realized for a variety of sizes and various efficiency levels, even if space is very restricted.

Just as for the swivel unit, mid positions can be defined for the translatory kinetic unit. The mid position is realized by a stop which is mounted at the side of the mini slide. The mid position can be adjusted infinetely variable and even be monitored by sensors. Thus the movement of the piston is stopped. If the switch cam is released by the stop cylinder, the piston can move to its end position. The mid position can be hydraulically dampened.

The above already explains the feature of kinetic devices with variable main function as compared to those with one defined main function. Optional positioning on one line requires a so-called linear axis, which can move to different mid positions with the help of a servo-drive.

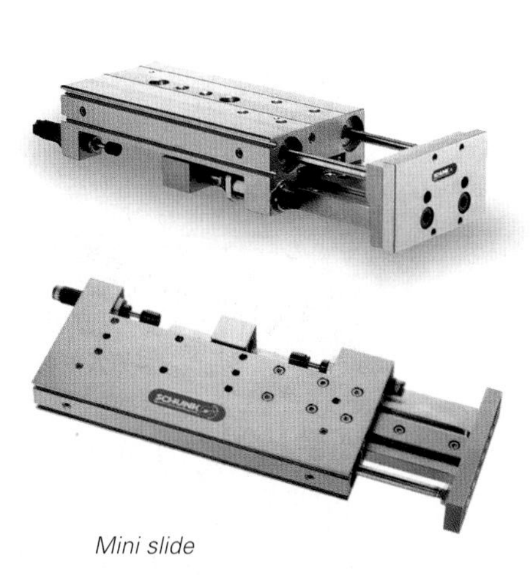

Mini slide

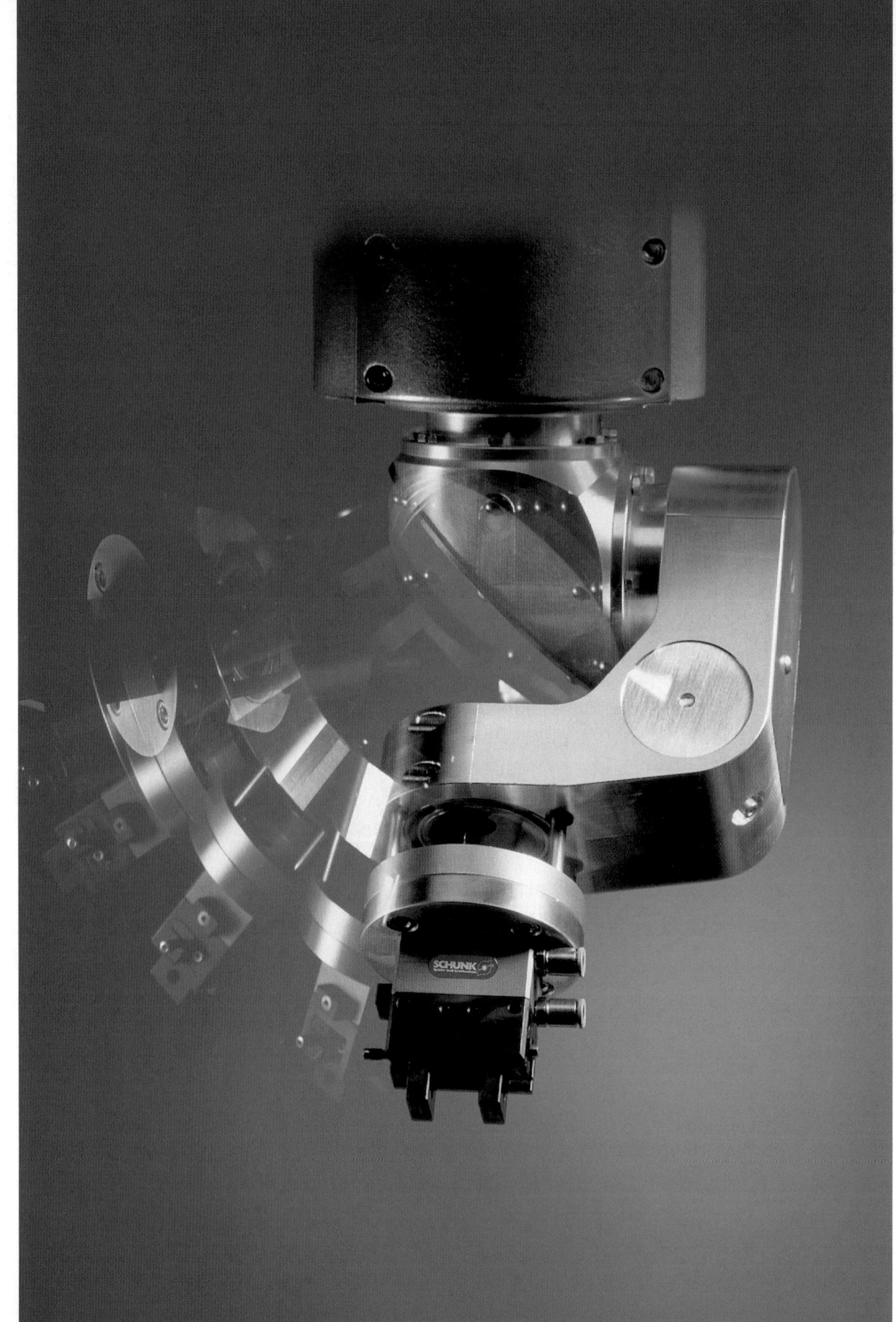
SCHUNK

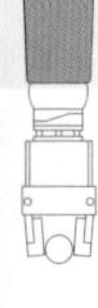

Free programming permits moving to any number of mid positions and their alignment on a line, a plane, or in space.

For a controlled movement, a system is required which coordinates drive, control, and measuring of the positions. The control of movement is the starting point. The movements, which are to be performed by the axis of the kinetic device, are defined by programming.

In the drive train depicted, the axis controller provides the servo-amplifier with the information necessary for actuating the servo-motor. The servo-motor then powers the transmission via a clutch and, subsequently, the robot or the linear axis. The axis controller receives a sensor feedback by the speedometer or encoder which is fitted to the servo-motor. Path measuring systems can be utilized, which provide the axis controller with information on the direct path of linear axes.

The aim of this construction is to keep the superior control free of control tasks and thus make the control circuit as fast as possible.

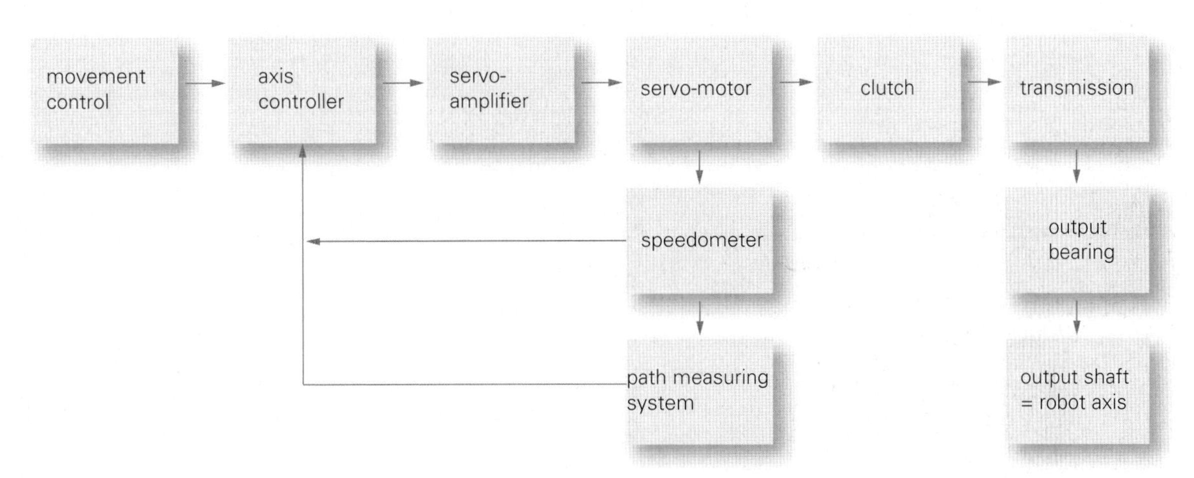

Figure 4.21 Components of a freely programmable axis

Linear systems:

linear gantry	*surface cantilever*
guided cantilever systems	*twin gantry systems*
quadriga	*quadro*

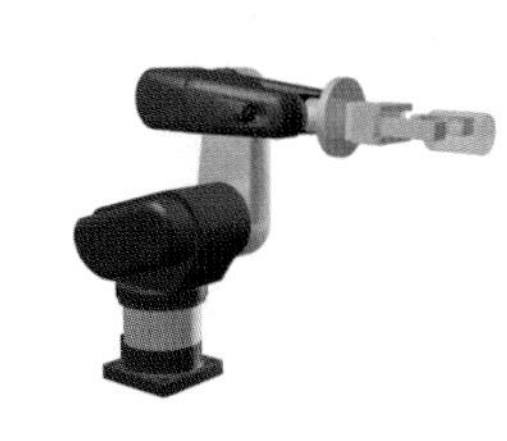

With the linear or rotary movements, it is possible to combine planes for reaching certain positions. If linear axes are combined in a three-dimensional system of axes, cubic workspaces are created. Each combination of axes results in a very specific workspace the workpieces can be moved within.

The main axes of the kinematics determine extension and shape of the workspace. With the help of the so-called hand axis, the robot can change the orientation of workpieces or tools. Workspaces of different sizes are created in relation to the respective axes length. This requires testing whether workpieces can be accessed and processing stations can be approached.

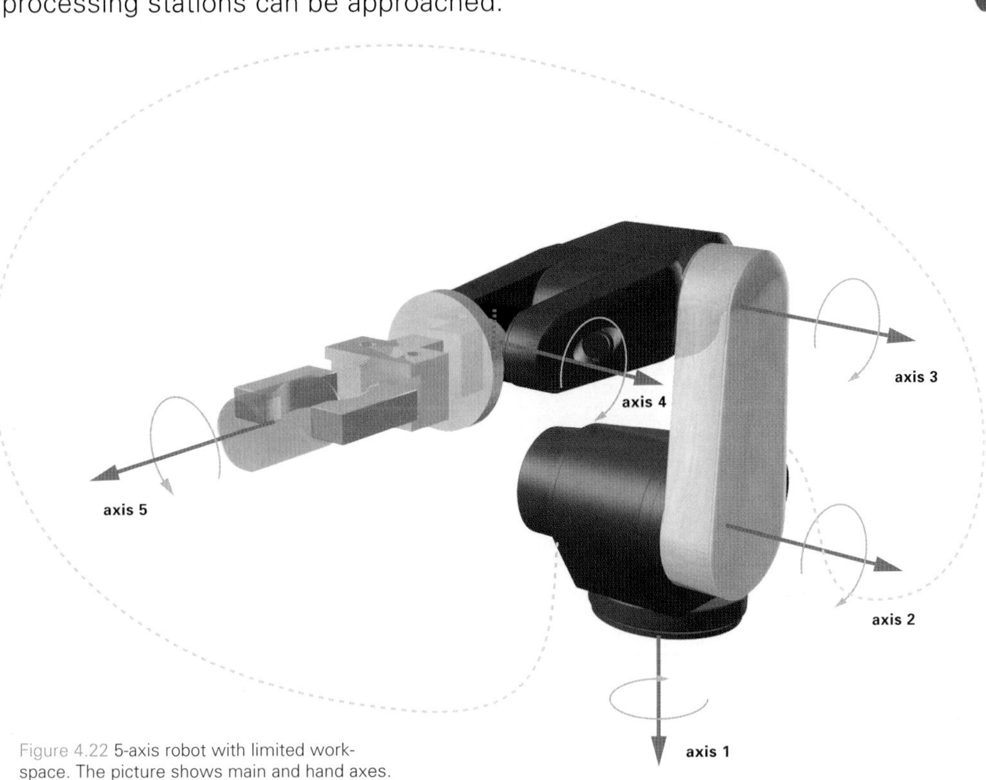

Figure 4.22 5-axis robot with limited workspace. The picture shows main and hand axes.

According to the mounting options, a kinematics` workspace can be used in different ways. The mounting options depicted can be found for nearly all types of kinematics, provided that the robot producer has included them, such as the option of mounting the robot to the ceiling.

Combinations of main axes		
Construction	Kinematics	Workspace

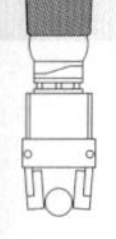

Table 4.4 Various workspaces of kinematic structures

Current kinematics are shown in table 4.4. By using translatory and rotary axes, differing workspaces can be created.

Parallel kinematics, which are exclusively used for being mounted to the ceiling, are an exception. They are specially designed to work above conveyor belts. This type of mounting saves space but usually causes static or dynamic problems.

Static calculations must take the weight of the robot into account. Considerable dynamic stress may arise from robot movements. As not every type of robot can be mounted to the ceiling, it is recommended that product specifications are carefully observed.

The workspace stated by producers is always calculated up to the hand flange of the robot. The hand flange is the part of a robot which a tool or a gripper is fitted to. The workspace of a robot is different from the workspace of a tool which is defined by the tool center point (TCP). Depending on the gripper design, the workspace of the tool can be very different from the workspace of the kinematics.

Any stress on the robot arm caused by the mass of the gripper/workpiece combination is summed up and defined as the payload. As shown in figure 4.22, the payload is already outside the workspace as stated by the producer. This inevitably leads to discrepancies between the ideal behavior of the robot without payload and the actual movement of the robot in a real operation.

Various mounting options for 5-axis robots – to ceiling, wall, and floor

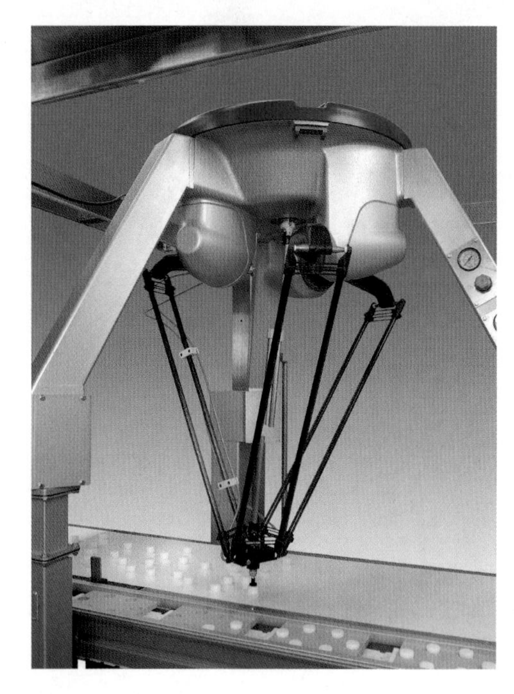

Parallel kinematics mounted to the ceiling by a specially designed frame (source: SIG Packsystems))

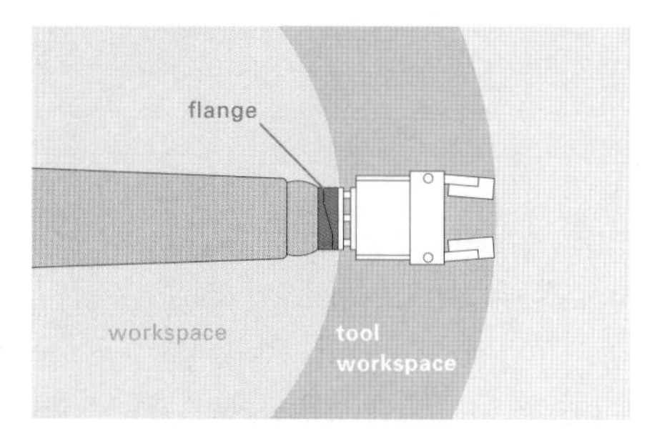

Figure 4.23 Restricted operational workspace and tool workspace

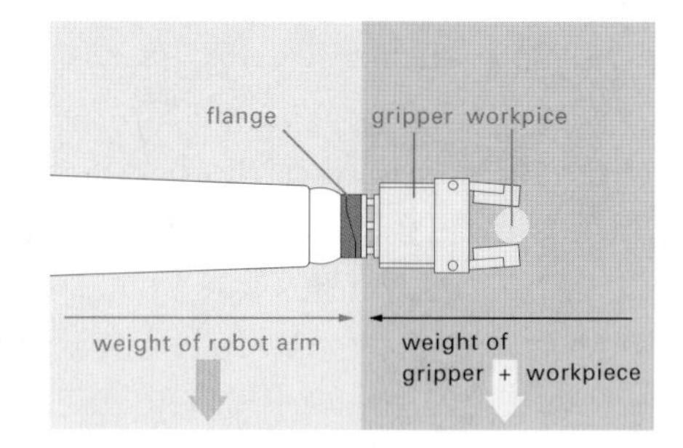

Figure 4.24 Stress on the kinetic device with gripper and workpiece combined

Producers offer so-called payload diagrams to help robot integrators to better estimate how the payload options of a robot change when the distance of the payload center of gravity to the robot flange changes. Figure 4.23 shows that differences in payload also occur in relation to the x- and y-direction of the hand flange depending on the construction of the robot`s hand axes.

	low payloads	medium payloads	high payloads	heavy payloads	special constructions
robot					
payload (kg)	0-15	15-60	60-210	up to 500	for special applications up to 500

Table 4.5 Various robots structured according to different payload categories (source: KUKA)

As shown in the diagram, the payload of the arm is reduced from 1kg to 0.5kg with a 80mm distance in the direction of the x-axis and a 40mm distance in the direction of the y-axis. These product specifications are stated with certain safety tolerances. Measuring or practical experience of the producer or the integrator is essential for critical applications.

The movements generated by the main axes are translatory or rotary movements. A velocity profile can be recorded for each movement, showing continuity or discontinuity of the axes` movement. Frequently, the maximum velocity of the axes is not reached because an acceleration phase needs to be interrupted by the following deceleration.

Figure 4.26 illustrates that the movement of a freely programmable kinetic device becomes more and more complex as the number of axes increases.

Kinematic axes configurations with more than three freely programmable axes are called a robot. A great range of robots which can be actuated with a standard robot control is available worldwide. Kinematics generate movement and, therefore, define the dynamic forces acting in relation to the workpiece`s weight and geometry.

The kinematics are covered by various tasks, as shown here by the differing workspaces of robots. This applies to both the shape of the workspace and the payloads to be handled by the robots.
The payloads result from the efficiency of the drive and the rigidity of the construction. The dynamics required by the task also influences payload and accuracy the workpieces are handled with.

Velocity profiles can be generated for each workspace of specific kinematics, which permits to make statements on how long the movement of a kinetic device will take.

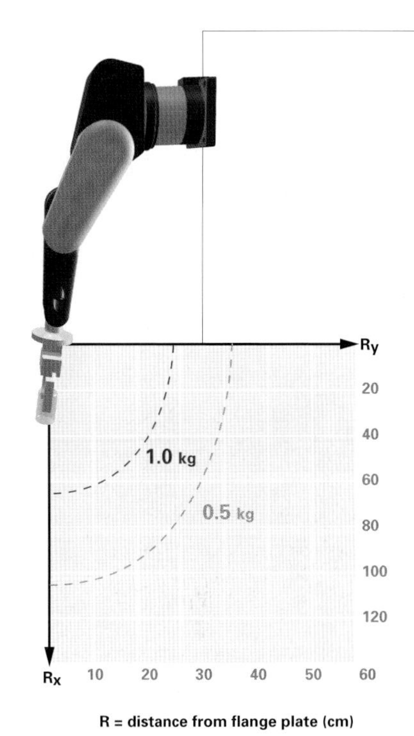

Figure 4.25 Example of a payload diagram for a robot with 1kg payload

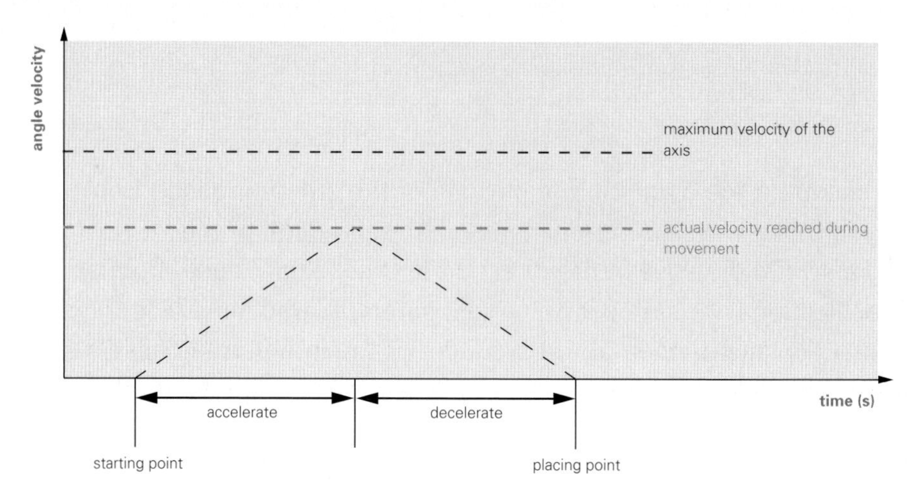

Figure 4.26 Velocity profile of a robot axis

Figure 4.27 shows the velocity profile of a linear axes system in the form of an x/y system of coordinates. The linear axes system is utilized for positioning workpieces below a processing machine. The workpieces can be positioned anywhere in the direction of the y-axis along a straight line of +/-300mm. The diagram in 4.27 is the result of measuring the duration of movement between the starting and placing point.

When using linear axes, the relation of movements is quite clear as the options for relating between starting and placing point are evident. If rotary axes with various levels of efficiency are employed, it becomes more difficult to relate to velocity. The main obstacle is the observer`s mental expectation to move the tool directly from A to B at the shortest distance possible. If the control is provided with a starting and placing point only, all axes will try to cover their distance as fast as possible. This fact is explained by the plan view of a SCARA robot.

If the robot is to move from P_1 to P_2, it must move two axes in the plane. However, a much larger distance must be covered by axis 1 than by axis 2, in this case 60° by axis 1 and only 20° by axis 2.
It cannot be assumed that the acceleration of the respective axes is always the same for a specific kinetic device or type of construction. As the movements and the constructions differ, the end position is reached at different points in time. This applies to both rotary and linear axes.

Therefore, the SCARA robot has a more complex velocity profile than a linear system. Cycle time can only be determined on the basis of exact knowledge of the distance between the respective starting and placing points in the workspace and the position of the robot in relation to these points.

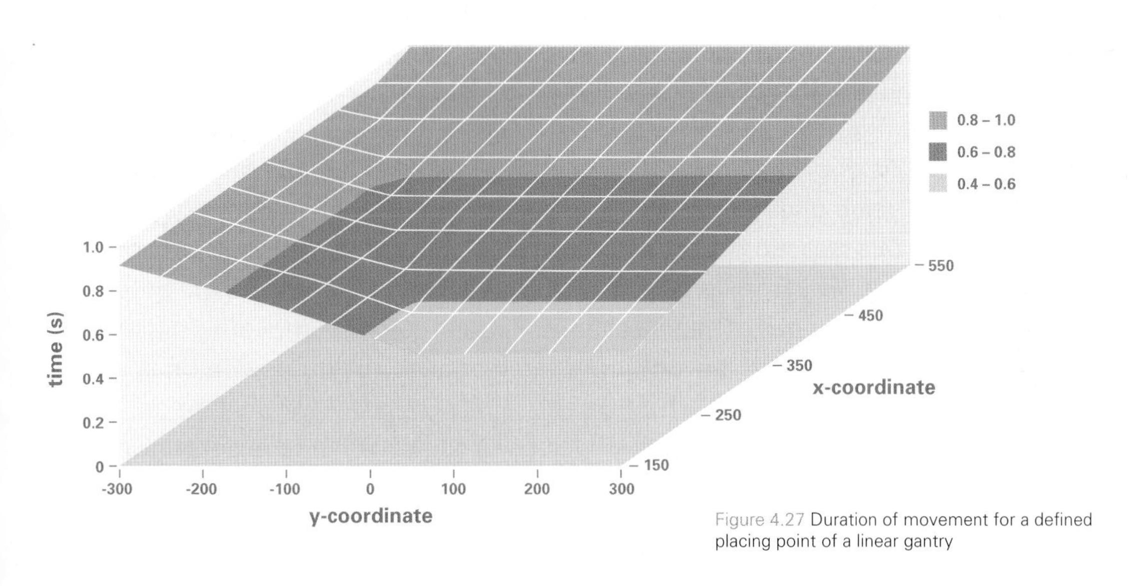

Figure 4.27 Duration of movement for a defined placing point of a linear gantry

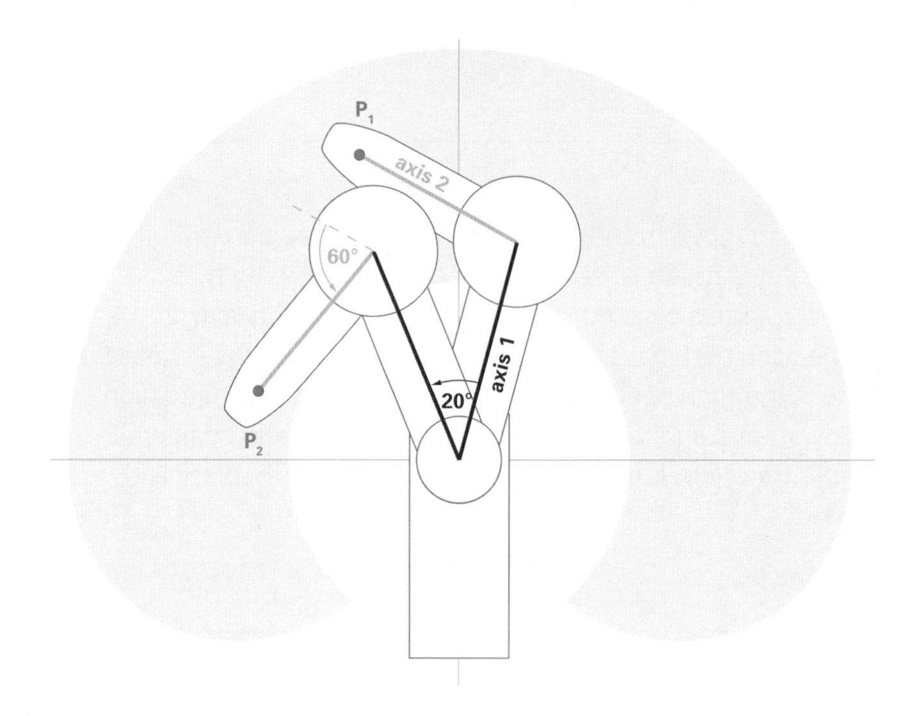

Figure 4.28 Robot movement subject to the condition that all axes are to reach their position as fast as possible

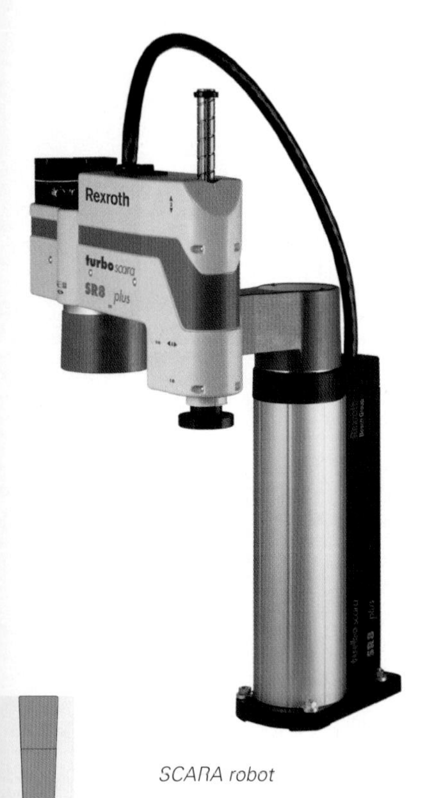

SCARA robot

Another example of influencing the cycle time of a SCARA robot`s movement becomes evident when looking at the distance covered by the tool of the robot. Axis 1 moves from 0° to 45°. Axis 2 does not need to change its position in relation to axis 1 in order to move to point P_2. Axis 2 must be moved only if point P_2 needs to be approached along a straight line.

As depicted, linear path control requires axis 2 to first turn in the same direction as axis 1 to reach point P_2, and then to turn against this direction in order to reach the placing point along a straight line. This makes overall movement decelerate as all axes must be coordinated. The so-called PTP (point-to-point) movement does not consider the movement of the workpiece or the gripper in space.

The axes approach their positions as defined by the control and thus reach their target.

The robot control must be provided with points of support to keep the movement of the workpiece or the gripper on a set path. The type of movement between these points of support is then defined as linear, for example. This definition of the type of movement is programmed and sets off the respective path interpolation within the control. Interpolation points are calculated, which guide the robot and its gripper along the path.

The points not only affect the three main axes of an industrial robot but also the three hand axes of a 6-axis robot. Hence the essential know-how of a robot producer consists of the calculations for converting the points of support into actual positioning points for the axes.

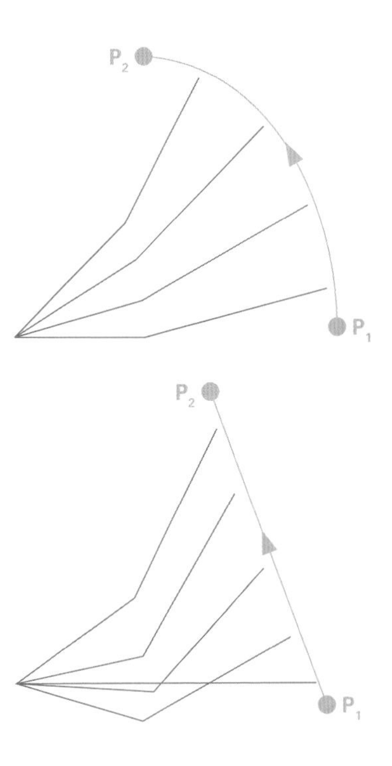

Figure 4.29 Movement under PTP control (above) and LIN path control (below)

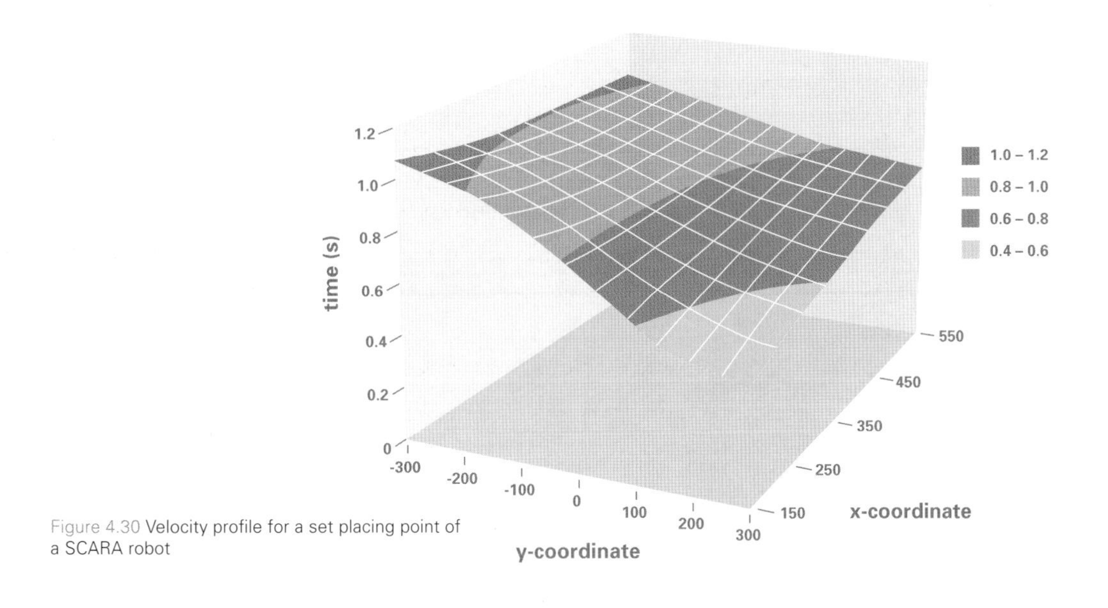

Figure 4.30 Velocity profile for a set placing point of a SCARA robot

The quality of movement performance always depends on how well the components, such as mechanical construction elements, drives, measuring system, and control, harmonize. The following performance characteristics mainly result from the mechanical system and the drives:

- radius of movement or workspace
- velocity of movement or velocity of the axes
- power of movement or payload

Together with the control and the measuring systems, they influence the following performance criteria:

- the flow of movement
- the precision of movement
- the continuity of movement or the accuracy of repeatability

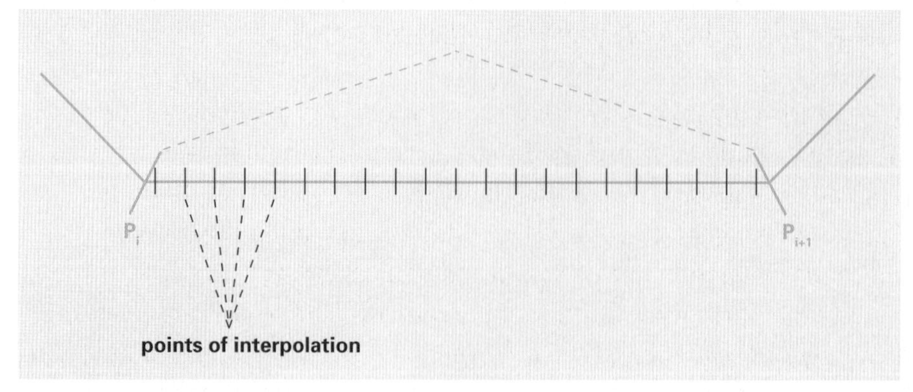

Figure 4.31 Section of a path with points of support and points of interpolation

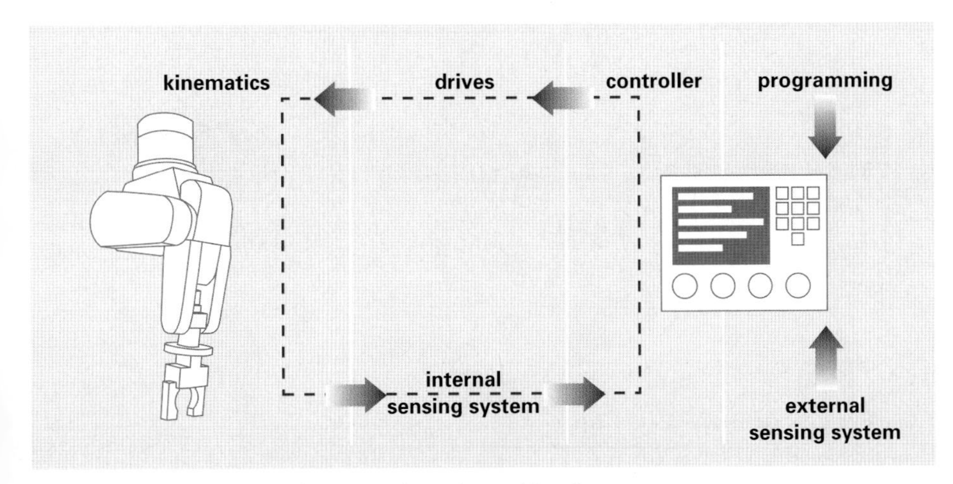

Figure 4.32 Core competence of a robot producer – harmonizing all components

Other important influencing factors include the type of programming and the influence of external sensors, which have to be taken into consideration when planning automated solutions.

If the robot is required to perform processing tasks on a workpiece, high demands on the flow of movement and the precision of movement are made. As explained, it is essential to move according to the path along the workpiece with the respective orientation.

Accurate repeatability is most important when approaching a point. The point needs to be reached as precisely as possible. Accuracy of positioning and accuracy of repeatability of a robot need to be distinguished. The latter is the result of a measuring series where the robot repeatedly moves from the same starting point to a measuring point. Compared to the accuracy of repeatability, the accuracy of positioning does not reach the same level of accuracy because the robot has to move from different starting points to a defined point in space.

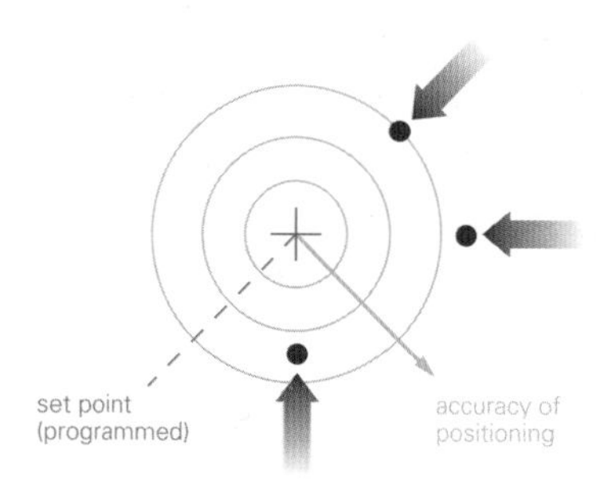

Figure 4.33 Deviations from the set point when approaching it from various directions (accuracy of positioning)

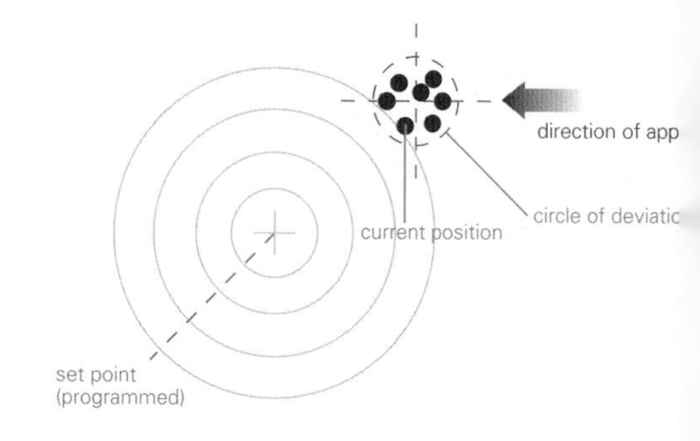

Figure 4.33 Deviations from the programmed set point when approaching from one direction (accuracy of repeatability)

The demands on the quality of a movement vary according to the moving task of the robot. In general high accuracy of repeatability can be assumed, which is sufficient for most applications.
For specially dynamic tasks in the packaging industry, the accuracy of repeatability is not essential because tolerances for the placing position may amount to several millimeters.

Placing in trays
(source: SIG Packsystems)

Robot producers offer simulation systems for a nearly realistic preview on the robot and its control in order to check the application option of different robot kinematics before the actual test.

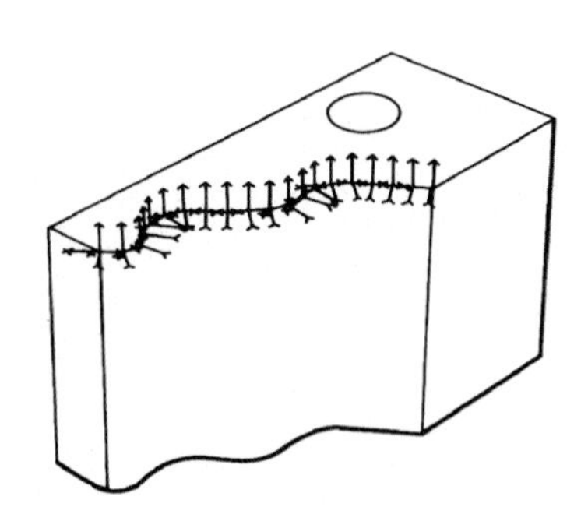

Robot path along a workpiece, generated from CAD data

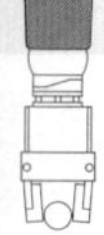

The gripper and the workpieces can be simulated to be able to depict and even program an application. Nevertheless, the ambient conditions which need to be taken into account for an automated solution are still hard to simulate, such as cleanroom or strict hygiene requirements.

The robots or gripper components have to be specially designed for the respective environment.

In the past few years, nearly all producers have endeavored to optimize automated solutions for the industries concerned. A broad range of application options is available for particular ambient conditions.

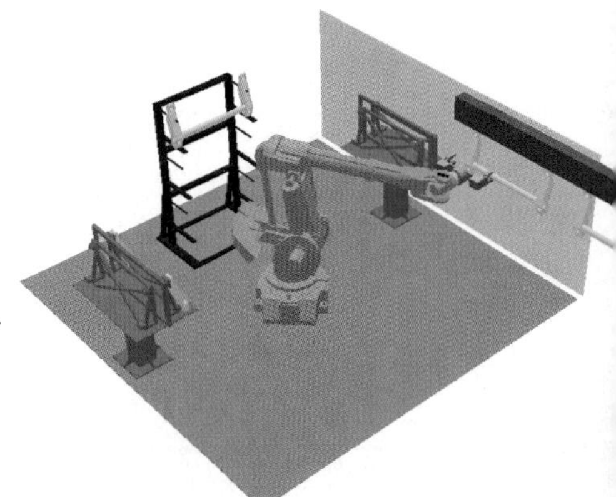

Simulation of a working cell

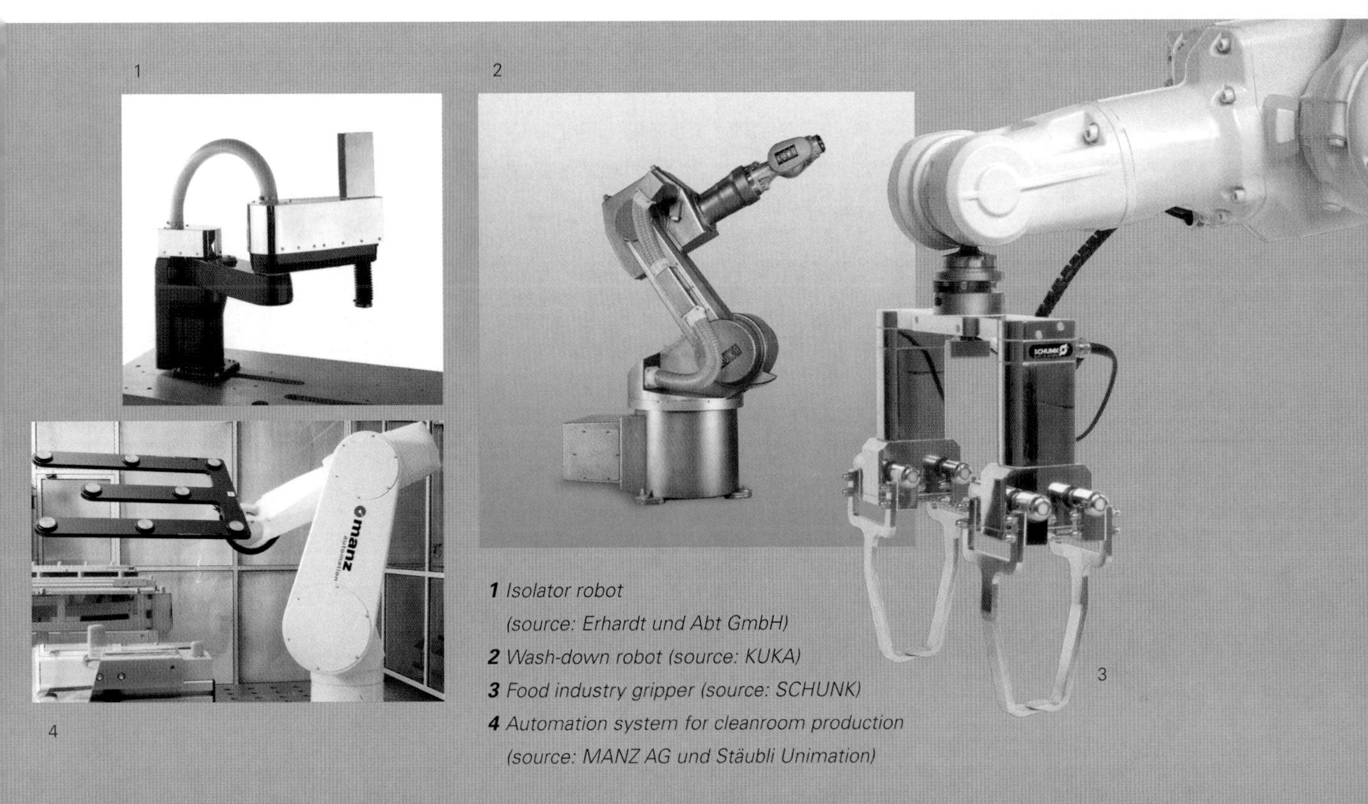

1 *Isolator robot (source: Erhardt und Abt GmbH)*
2 *Wash-down robot (source: KUKA)*
3 *Food industry gripper (source: SCHUNK)*
4 *Automation system for cleanroom production (source: MANZ AG und Stäubli Unimation)*

5.8
5.3
5.11
5.2
5.17
5.13
Dole
MOTOMAN robotec
5.1
5.7
5.9
STÄUBLI
5.4
5.17
5.10
5.14
GALAXY
5.12
5.1
Erhard+Abt

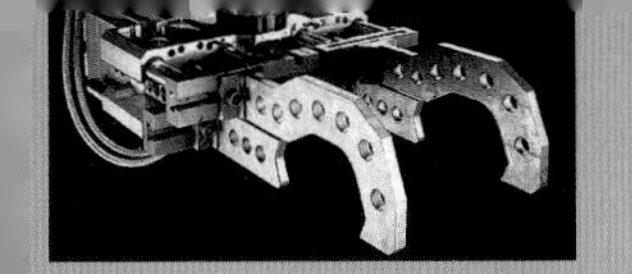

5.18

5.5

5.5

5.16

5.6

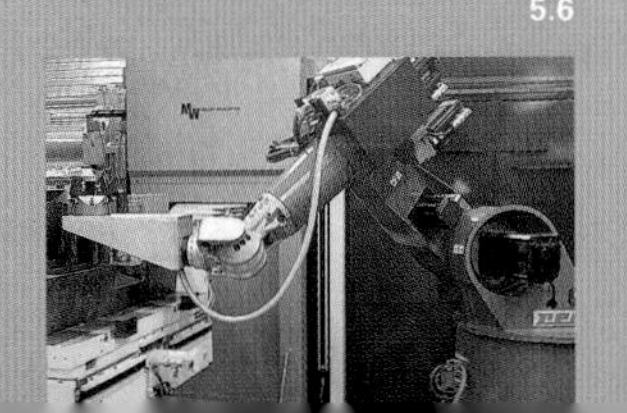

Within the volume of this book we can only include a small part of applications from the vast amount of documentation and photos we received. At this point we would like to thank all who provided us with material.
Of course, we are interested in more case studies with photos. Please send your material to the following address:
book@robomotion.de
We also appreciate your suggestions on the contents of this book to the same address.

5 Practical Applications

5.1 Precision and Coordination

Some sensitivity and up to three smoothly operating hands are needed for delicate operations. The robots by Motoman (Yaskawa) fulfill these requirements as you can see from the pictures of their trade fair presentation. The three planet wheels can only be mounted together with the central drive gear, while several robot hands move at the same time.

Welding bicycle frames is quite a similar task. Two tubular workpieces are positioned by a robot hand each, the third hand operates the welding gun. The axes of the robots handling the workpieces are synchronously moved to maintain the welding pool in a horizontal position.

Manufacturing reversible blades made of hard metal requires most precise handling. The photo of Manz AG (right, large fig.) shows the revolver gripper head of a robot handling the highly precise workpieces in a very gentle way.

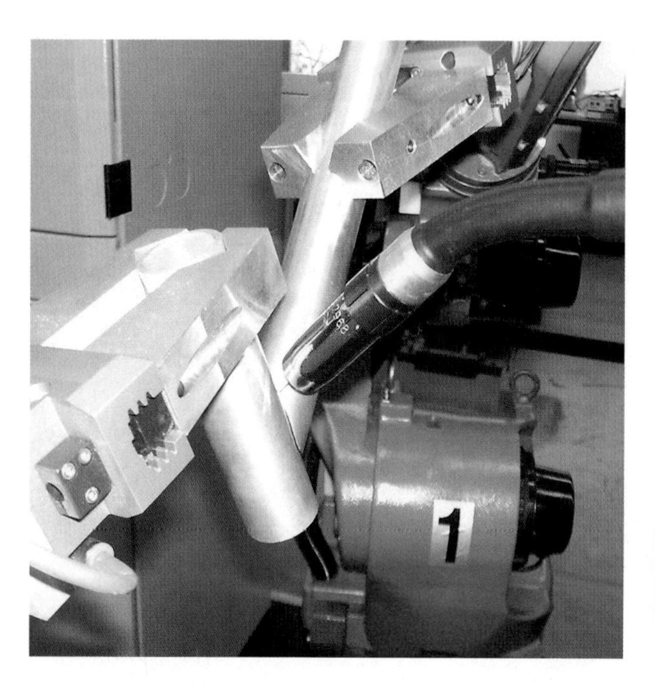

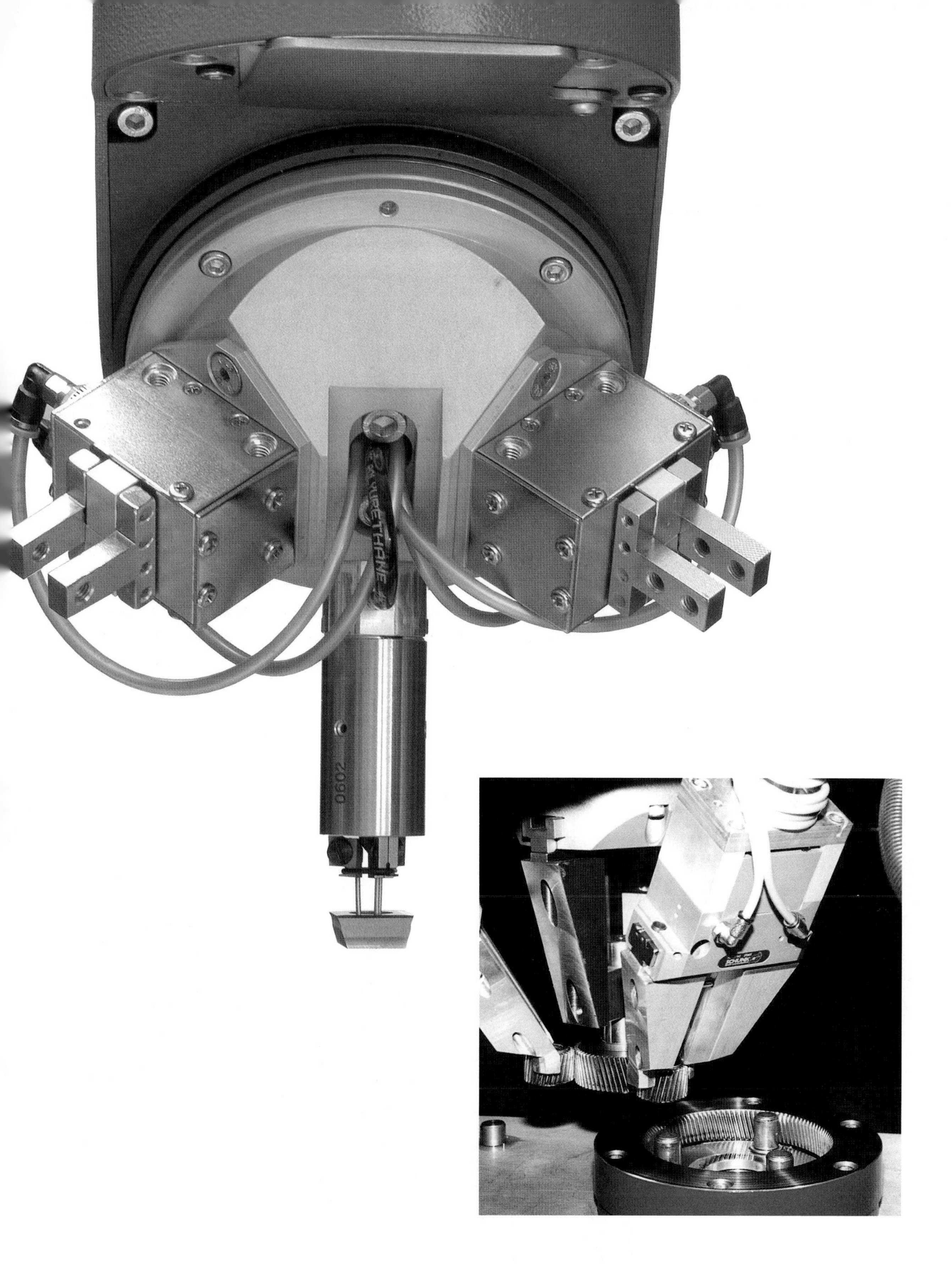

5.2 Velocity

When we are talking about velocity, we are talking 0.07 seconds per workpiece or 50,000 workpieces per hour. Siemens Production and Logistics Systems AG in Munich, Germany, work with four gantry systems equipped with a revolver head each and an average cycle time of 0.28 seconds per handling operation.

Each revolver head can take up to twelve gripper or pipette modules. The pipettes are designed as suction grippers. If workpieces are handled which cannot be safely gripped by a pipette, angular grippers by SCHUNK are employed. The gripper is opened by spring force because clamping at the revolver head is only possible by vacuum (-0.65 bar). It is closed by a membrane converting the vacuum into a closing movement. Thus gripping forces up to 5N are achieved at a gripper weight of less than 20g. Even workpieces without a smooth surface can be gripped safely and fast by vacuum suction, such as protection frames for mobile phones, Sub-D plugs, and more.

5.3 Color Sorting

Despite instructions for workpiece assembly, errors may occur if they are not quite clear or if workpieces are not allocated accordingly. In our example four tubes of the same diameter have to mounted to an automobile filter.

The manufacturer decided to use color codes to prevent the tubes from getting mixed up. The color code is realized by colored synthetic rings which are monitored and then isolated by a color sensor before set-up. The robot by Erhardt & Abt in Kuchen, Germany, is provided with the rings for each connecting piece at the filter and then mounted. As a result, the operator is able to match the tubes and the connecting filters without any doubt. The colored rings are removed after assembly.

5.4 Big and Heavy

DaimlerChrysler produces heavy-duty axes in their subsidiary in Gaggenau, Germany. The robust workpieces are mainly built into heavy trucks or used for long-distance traffic. High demand and increasing cost pressure make an efficient automated solution indispensable.

Formerly, three milling machines, two broaching machines, and two trimming or cleaning machines were used in the manufacturing process. Subsequently, the milling machines were to be operated by a flexible and fast robot. The robot is placed in between the three milling machines, grabs a milled workpiece from the transport system, and feeds it to a milling machine. The milled workpiece is

taken out by the two-finger parallel gripper. Then the three-finger concentric gripper PZN 200 transfers the workpiece, the internal diameter of which has not been processed, to the lathe chuck. The milled workpiece leaves the robot cell on a conveyor. Samples are taken from the conveyor by hand to check shape and dimensional stability.

Two other systems are integrated into the automated process for trimming and cleaning the workpieces. The next step is the use of another robot to palletize the workpieces for further transport.

The team of the PGP 4 department at DaimlerChrysler in Gaggenau, Germany, realized this project despite an extremely short project phase. The simplified working process is highly appreciated by the operating staff.

5.5 Wheel and Wheel Rim Handling

Wheel rims are so-called bulk goods, i .e. thousands of them are produced every day. In order to transport the wheel rims forth and back between a conveyor and a processing station, Stahlschmidt & Maiworm uses a KUKA robot in combination with a SCHUNK dual gripper.

The dual gripper reduces cycle time as the robot takes a workpiece, which is due for processing, to the processing station and picks up the finished workpiece. The large-scale dimensions of the wheel rims do not require a swivel unit, but the robot makes a move to position the new workpiece. The wheel rim gripper is equipped with a large stroke and is able to cope with most different diameters.

The gripper taking worn out tires to the reycling system (fig. left) does not need to be very sophisticated. Gripping the wheel and placing it on the conveyor with an accuracy in the centimeter range is more than sufficient.

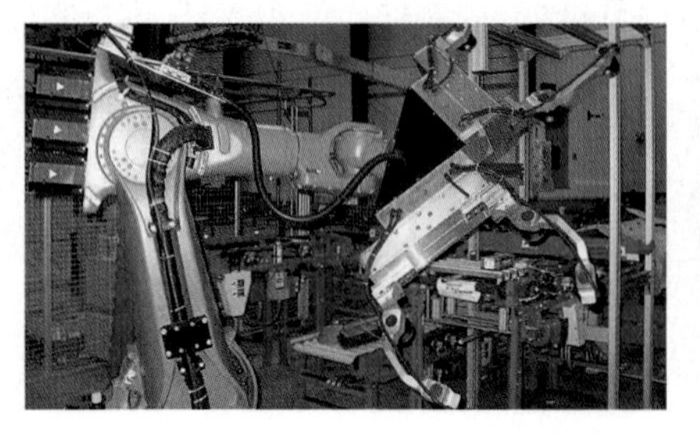

5.6 Heavy and Delicate

Aluminum rolls with a 40cm diameter weigh an impressive 50kg. In our example, it was not possible to pick up the rolls from the outside because the workpieces had to be taken up from a pallet. Therefore, the rolls are picked up by a triple concentric gripper from the inside opening. The gripper fingers were designed to maintain the weight of the rolls by either force-fit or form-fit gripping.

Thus the four-axis robot can pick and place the rolls from the pallet as well as from the cutting machine. In addition, increased air pressure is used to ensure the necessary gripping force. Two rolls are taken up in one stroke, the gripper needs to bring both gripping components into the right distance to each other. A spindle driven linear axis takes the gripper into position before it enters the opening of the rolls. At the same time, an optical sensor monitors the position of the workpiece. The robot service R2 GmbH in Rodgau, Germany, realized this system and built the gripper on the basis of SCHUNK components.

5.7 Lightweight

Packaging designers are highly creative, especially if packaging with an individual look is on demand. As a result, packaging machine producers have to meet this challenge with suitable solutions. SIG Packsystems AG in Schaffhausen, Germany, comes up with most innovative solutions for the clients with the most demands.

In our example, a special "swivel unit" mounted to the suction gripper of the robot enables the cookie to be picked up horizontally from a conveyor and, subsequently, be placed into the packaging machine at a 45° angle. The fourth axis of the robot works as a drive for the special swivel unit because it was not possible to cope with the weight of a separate drive.

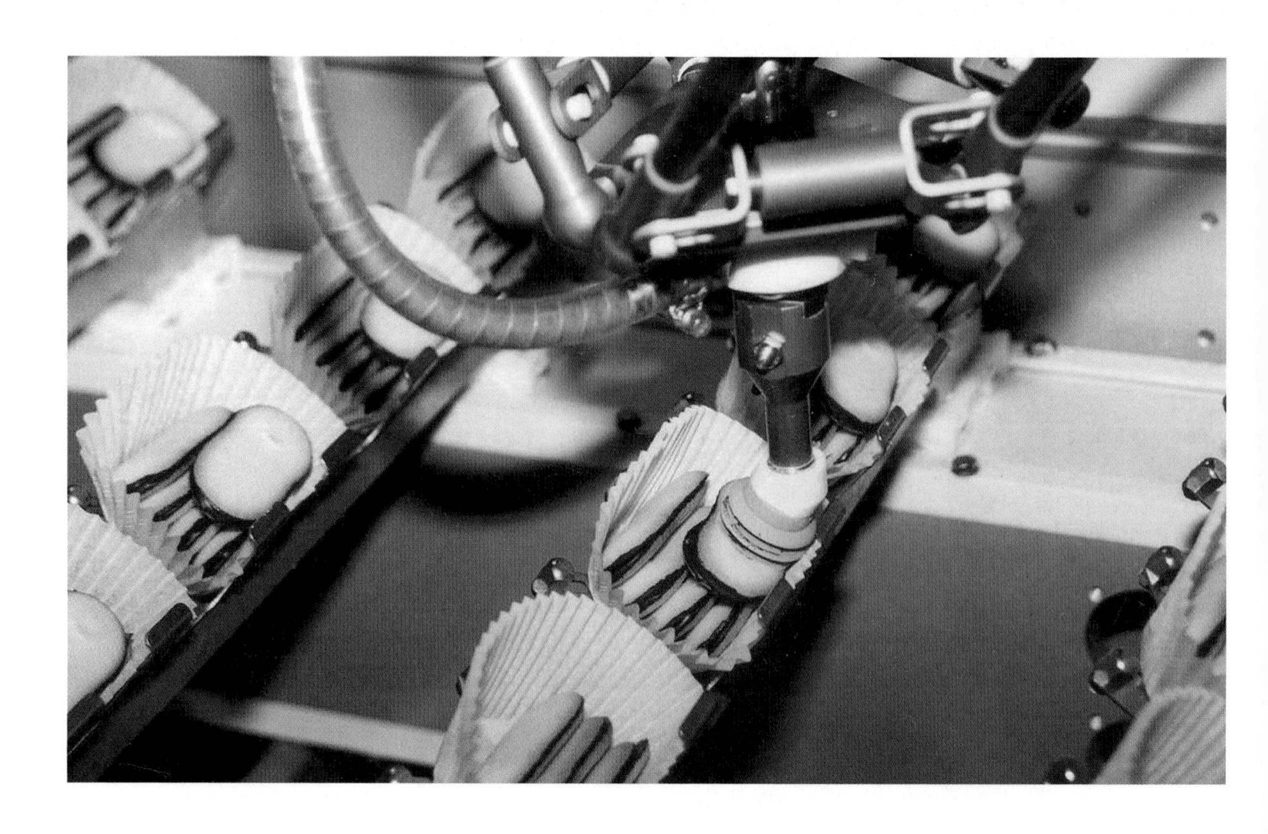

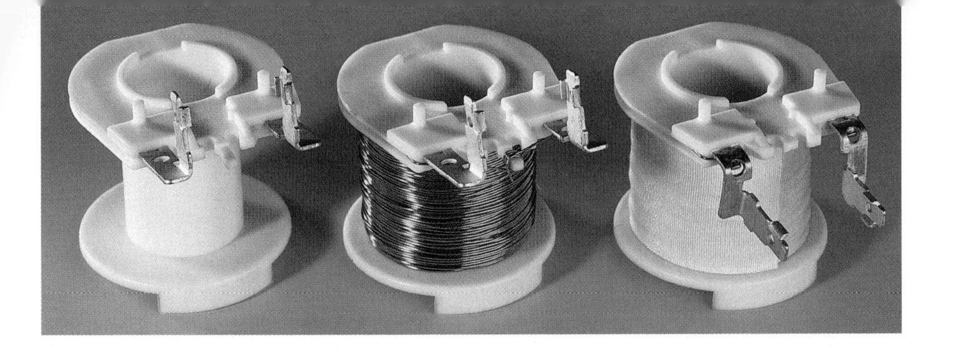

5.8 Tooling Machines

The Swiss company Meteor Maschinen AG is specialized in winding machines for producing expensive electrical coils. Robots by Bosch Rexroth AG coordinate the single steps of production and permit an extension of the production process due to their high degree of mobility and flexibility.

As a link between workpiece feeding, winding machine, and production platform, the swivel arm robots works with single-actuated gripper modules, which can be programmed each according to the respective task. Large degrees of freedom in relation to the pick- and place positions facilitate a flexible integration into diverse production processes. The robots are either used as standard modules or changed over and programmed for different tasks in no time. The control of these modular concepts remains the same, which results in similar, user-friendly interfaces for different tasks.

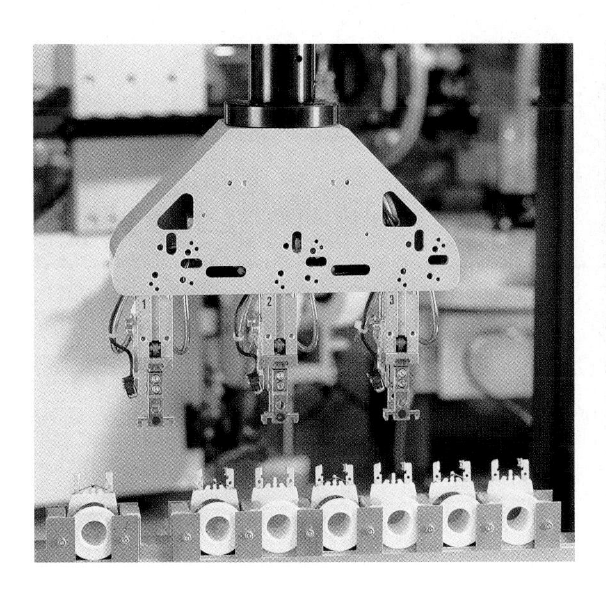

5.9 Dirty Environments

Robots and their grippers must be very robust in this kind of environment. Pictures of KUKA Robotics GmbH in Augsburg, Germany, show their robot in two filing applications. No matter if you are filing chain saws or household knives: The filing residues remain the worst enemy of automation technology. Not to forget the stress of process forces during filing, while robots and grippers are expected to work as maintenance-free as possible. Pneumatic actuation can prevent minute dust particles from entering into the gripper.

Erhardt und Abt in Kuchen, Germany, had to deal with the problem of using a coolant while feeding processing machines. Similar to the robust gripper, the camera was equipped with an extra housing to protect it from the effects of the coolant.

5.10 Cleanroom

Coating modules for production of solar collectors requires highest purity and precision, and had always been a typical example of purely manual handling. Recently, this type of cleanroom application was taken over by a robot. The automation system for cleanroom production by Stäubli Unimation was equipped with a special gripper system which can take in twelve modules and position them with high precision in the cleanroom. As a result, more modules can be placed on the carrier systems. This equals an increase in efficiency by 44 percent per coating. The system solution was realized by Jonas & Redmann GmbH. Manz AG in Reutlingen, Germany, has specialized in robots handling thin glass plates which are used for the production of LCD monitors. The figure on the right (below) shows such a robot with its special gripper.

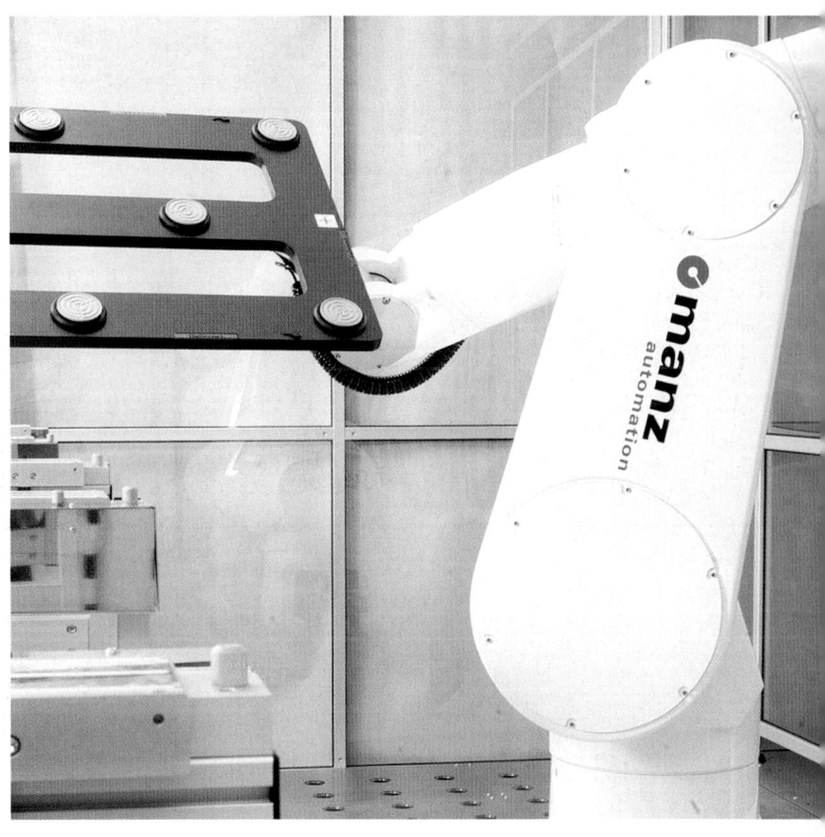

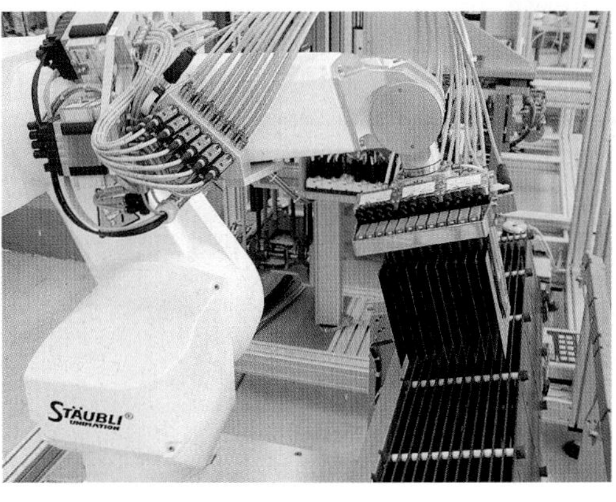

5.11 High Temperature

An ambient temperature of up to 750°C is not suitable for human fingers. Yet, even robots have their difficulties with these temperatures. The robot developed by Reis Robotics must be constantly cooled while taking samples out of a melting pot. Cooled with a blast engine, the robot is able to remain near the melting pot for up to three minutes to take the sample. As you can see from the picture, the robot is stressed by the temperature as well as the dust in the air around it.

5.12 X-Ray

A special challenge for grippers is x-raying workpieces. This is a typical example of a robot application in environments too dangerous for humans. YXLON International X-Ray GmbH, Hamburg, offers non-destructive material testing on the basis of X-ray technology.

A practical application for an end user in the automotive industry included testing aluminum castings for pores or shrinkholes with the help of X-rays. This automated check is indispensable in terms of safety aspects, especially for supporting chassis parts.

If the service life of workpieces needs to be reliably determined, X-ray technology is a very interesting option. The gripper fingers have to be designed in accordance with the X-ray to avoid faulty measuring.

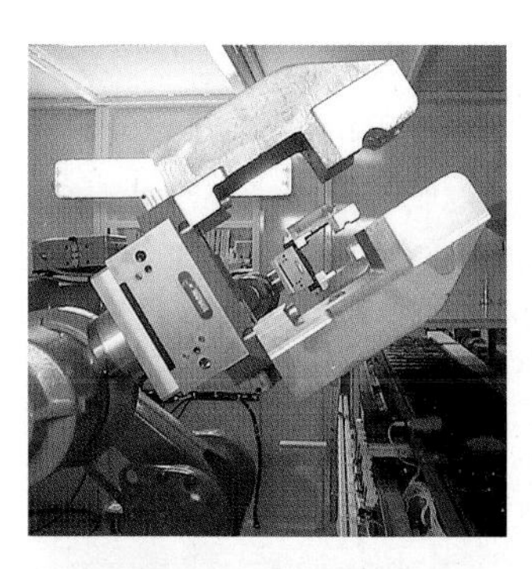

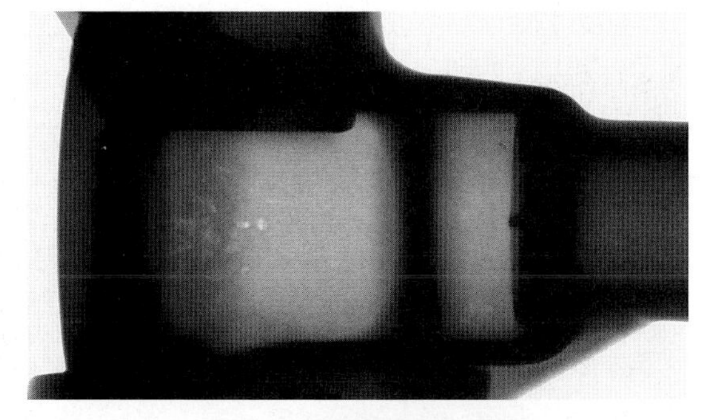

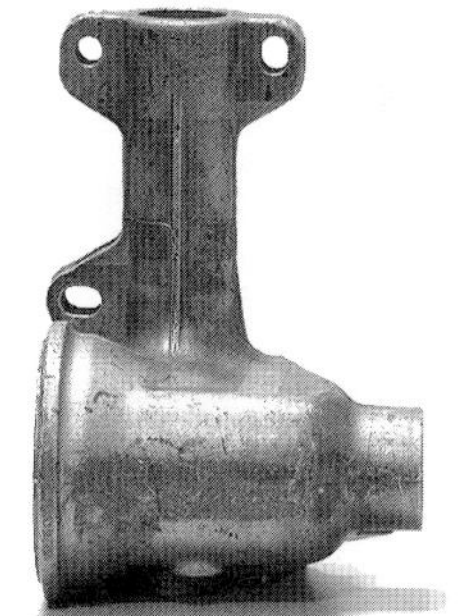

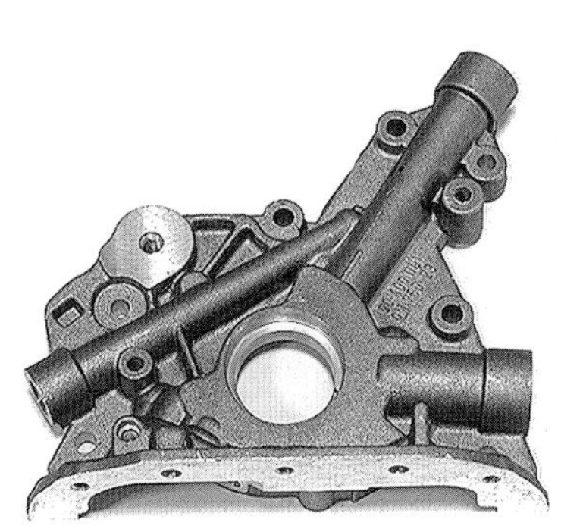

5.13 Logistics / Bananas

Before the bananas of the Coop Switzerland are delivered, the quality of the fruit and the pallet size are monitored. The specific requirement of this task is gentle handling, i. e. the cases with the delicate fruit must be moved softly and protected from tilting. The automated system must also be capable of handling cases, which have been pressed against each other or arrived in a sloping position.

As a result, kink-arm robots had to be employed. The Coop Switzerland decided on three long-range palletizers KR 160 PA. The five-axis KUKA robots reach up to six pallet spots, and their special gripper systems act in relation to the specific task on hand, whether it be gentle or robust. Within one position, the gripper can also compensate for tolerances caused by cases that have slipped. Displaced stacks are compensated by a centering frame and a floating supported chain conveyor.

5.14 Hygiene / Meat

The gripper had to be able to pick up a pork loin, for example, and place it into a food processing machine. A payload of 6kg and integrated sensors for monitoring the gripping process were demanded. However, the task was not limited to this. The customer also demanded strict hygiene measures to avoid germ formation on the gripper or the gripper fingers.

The latter can only be achieved by a specific gripper design and regular cleaning during the handling process. The food industry gripper was developed by SCHUNK and robomotion GmbH in a joint effort. Sealed to IP67 requirements and an easy-to-clean design, a special type of food industry gripper was developed, which covers the most differing workpieces and great tolerances with an opening angle of 20° to 180°.

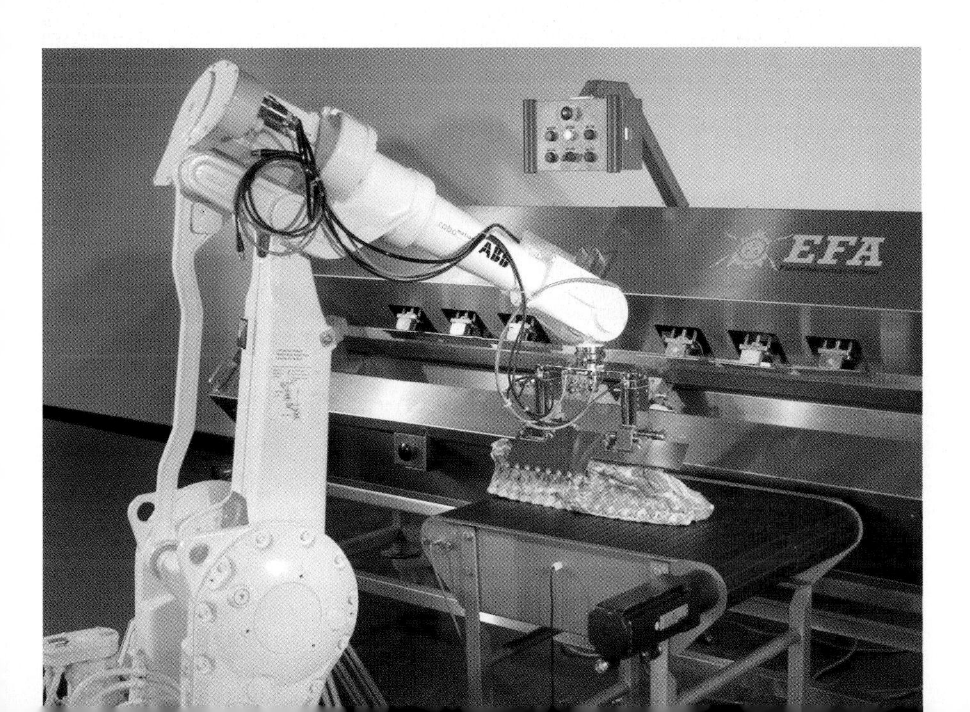

5.15 Cleaning / Cheese

ALPMA GmbH in Rott am Inn, Germany, realized an automated solution for handling cheese by using SCHUNK DWG modules. Standardized solutions for cheese manufacturers need to fulfill requirements of cleaning and agressive ambient conditions caused by salt solutions. With a minimum two shift operation, the systems must be regularly cleaned.

As illustrated, the gripper picks up the cheese portions from a conveyor. As not all of the cheese portions arrive in the appropriate position, two swivel units are integrated into the gripper to turn the cheese bits 90° before inserting them in the packaging machine. A quadruple gripper is necessary to achieve a packaging efficiency of about five metric tons per hour. Strict hygiene requirements make short cleaning intervals necessary. A gripper change system permits to separate the gripper from the robot within seconds and clean the gripper in a cleaning bath. The sealed angular gripper DWG has been working trouble-free for two years now.

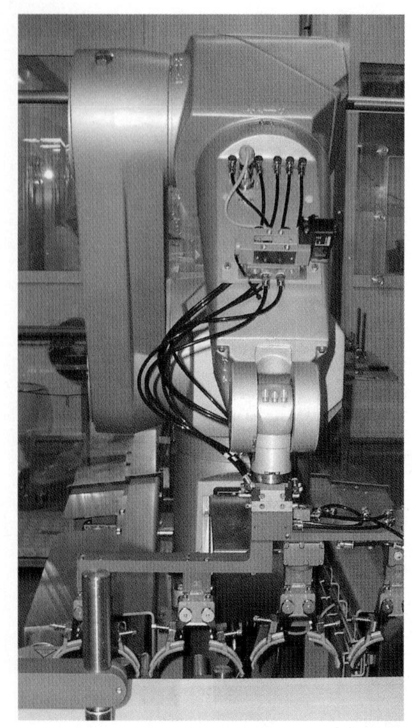

5.16 Barrels

Cylindrical workpieces of a special kind are barrels which are handled by a robot of ABB Friedberg in Germany. The multifunctional system copes with several tasks.

In general, closing the barrels with a cover and placing them on a pallet are two separate tasks which the gripper must cope with. In other words, it is a handling process and an insertion process carried out with one and the same tool.

Before the barrel can be placed on the pallet, the robot must get an empty pallet from the stack. The latter is another task the robot needs to cope with. The workpieces can hardly be more diverse than in this example, making the gripper multifunctional in the true sense.

5.17 Beverages

A "beverage robot" of a special kind works with a flexible gripper and the appropriate robot technology. The gripper must be able to handle all available cases, ranging from the smallest to the biggest sizes. Reis Obernburg has developed a flexible gripper which adapts to the various case sizes. The number of beverage types is limited only by the storing space in reach of the gripper. Two compact axes are used which cover up to 15m pallet store. The gripper is even able to pick up cases in narrow spaces, for example, if they are surrounded by high stacks of other cases, as depicted above. Combined with a simple human-machine interface, the service robot can be operated by anyone and is available around the clock, just like an ATM. ABB concentrated on de-stacking whole layers of beer cases, a strong-man act even for a gripper.

5.18 Fast Snacks

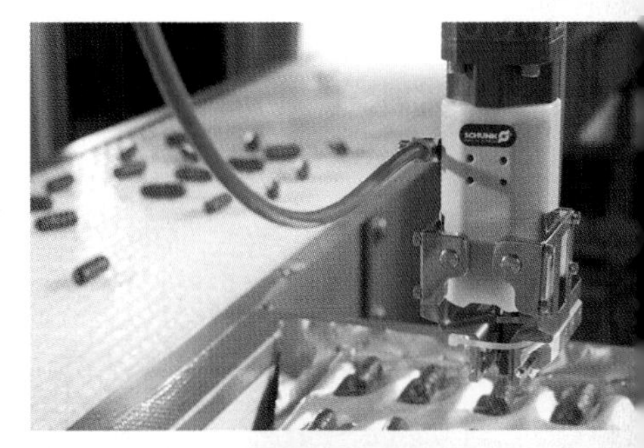

Consumer goods are fast running products in the market. And the markets change fast and faster. Hence, if a product becomes a blockbuster the industry has to cope hard with the production to get the amount of products they could sell. On the other hand, the product has to be changed in many ways in order to keep it in the market and to earn the money which has been spent in the development phase.

Often a change of a product comes in the form or the shape of the packaging style. Robots are highly flexible and they are already fast – combined with a special gripper technology they work "high speed". The shown example is a combination of a high speed Flexpicker and a high speed mechanical gripper designed by robomotion and SCHUNK both from Germany. The shown principle can handle easily up to 160 products a minute in this application with snack sausages. More you can eat per minute!

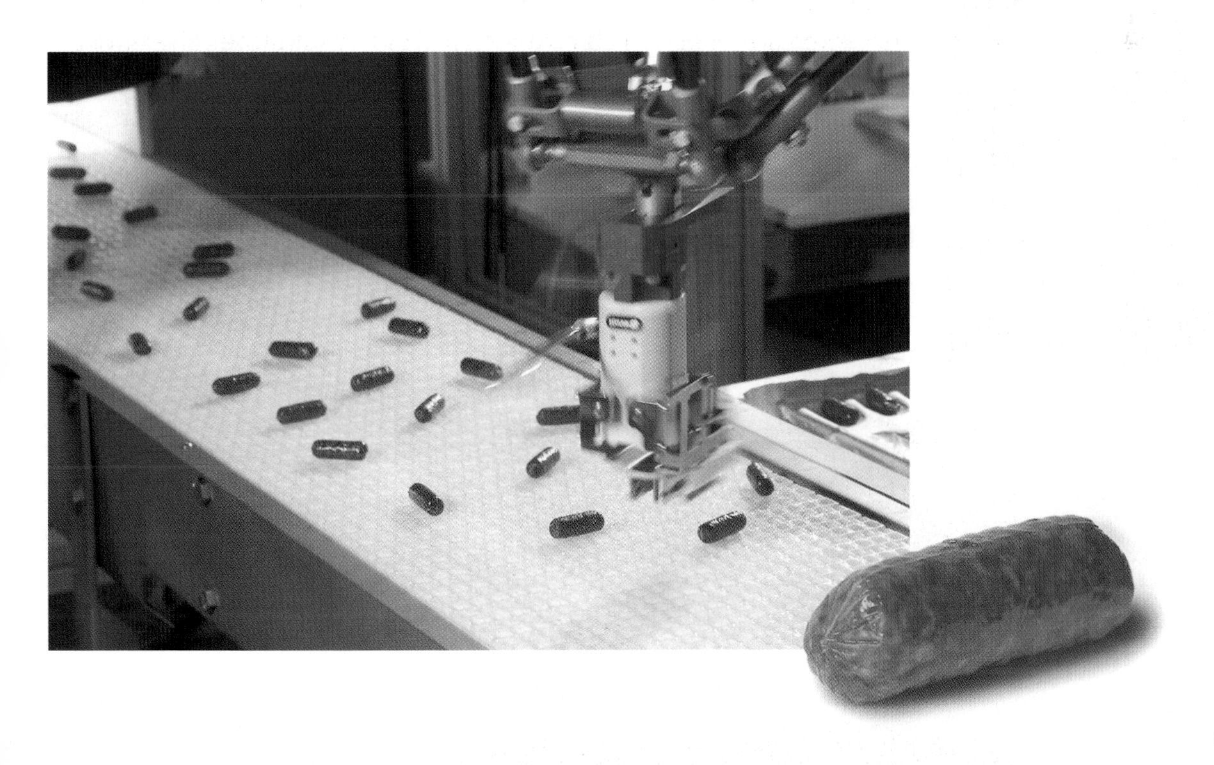

5.19 Dairy Industry

Robots in agriculture may still be a bit far from our imagination but they are already working successfully and efficiently. The Galaxy milking robot permits the animals to control their milking intervals themselves. A floodgate automatically opens, tempts the animals with some food, and recognizes them individually. A soon as the cow has reached the position within the milking box, the milking process starts with one robot arm optically sensoring the udder and positioning the milking machine. The udder is cleaned and milked in one step. The amount of milk is documented and the cow is let out of the milking box.

The milking intervals are selected freely and are more animal-friendly than regular milking at set hours. Apart from the substantially lower work input, the system is supposed to produce a higher yield. Insentec B. V. in the Netherlands offers the system under the product name Galaxy.

5.20 Greenhouse Production

It is increasingly difficult to find staff for the hard physical work in greenhouses. A company like ISO Groep in the Netherlands, turned the back on its former key competence of managing greenhouses and ventured into mechanical engineering to develop automated solutions for horticulture.

Up to 42 seedlings per stroke can be handled by the SCHUNK grippers. The robot copes with an impressive 168 seedlings per minute which requires a minimum 2.5ha greenhouse floorspace to ensure the efficient use of the robot.

Obviously, there are enough greenhouses of this size interested in the new technology. According to ISO Groep, they are fully booked until 2007 – another example of a new idea turning into a successful automated solution.

- Bark, Karl-Borries:
  Adhäsives Greifen von kleinen Teilen mittels niedrigviskoser Flüssigkeiten.
  Berlin u.a. : Springer, 1999
  (IPA-IAO Forschung und Praxis 286). Stuttgart, Univ., Fak. Konstruktions- und Fertigungstechnik, Inst. für Industrielle Fertigung und Fabrikbetrieb, Diss. 1998

  Biagiotti L., Lotti G., Melchiorri C., Vassura G.
  Design Aspects For Advanced Robot Hands
  IEEE/RSJ International Conference, Lausanne, 2002

- Bräuning, Uwe:
  Pakete und Behälter automatisch im Griff: Pilotprojekt: Roboter erobert neue Einsatzgebiete in der Postautomation.
  In: Materialfluss 32 (2001), Nr. 4, S. 62-63

- Hesse, S.
  Automatisieren mit Know-how : Handhabung, Robotik, Montage.
  Darmstadt : Hoppenstedt Bonnier, 2002

- Hesse, S.
  Greifer-Praxis, Greifer in der Handhabungstechnik
  Vogel, Würzburg, 1991

- Kallweit, W.-A.
  Miniaturgreifer nach biologischem Vorbild
  Dresden, 1988

- Kunstmann, Christian:
  Handhabungssystem mit optimierter Mensch-Maschine-Schnittstelle für die Mikromontage.
  Düsseldorf : VDI Verlag, 1999
  (Fortschritt-Berichte VDI: Reihe 8 751). Darmstadt, Techn. Univ., Institut für Elektromechanische Konstruktion, Diss.1998

- Meinel, K., Schnabel, G.
  Bewegungslehre, Theorie der sportlichen Motorik
  Sportverlag Berlin, 1998

- Menzel P., Faith D'Aluisio
  Robosapiens, Evolution of a new species
  MIT Press, Cambridge, Massachusetts, 2000

- Nicolaisen, P.
  Sicherheitseinrichtungen für automatisierte Fertigungssysteme Hanser Verlag München, 1993

- Rohrmoser, B.; Parlitz, C.
  Implementierung einer Bewegungsplanung für einen Roboterarm.
  In: Dillmann, Rüdiger (Tagungsleitung) u.a.; VDI/VDE-Gesellschaft Mess- und Automatisierungstechnik (GMA) u.a.:

- Robotik 2002 : **Leistungsstand, Anwendungen, Visionen, Trends.**
  Tagung Ludwigsburg, 19. und 20. Juni 2002.
  Düsseldorf : VDI Verlag, 2002, S. 59-64
  (VDI-Berichte 1679).

- Seegräber, L.
  Greifsysteme für Montage, Handhabung und Industrieroboter
  Expert Verlag, Ehningen 1993

- Warnecke, H.-J.
  Betrachtungen zur Automatisierung der Handhabung
  3. Internationale Symposium über Industrieroboter, Zürich, 1973

- Warnecke, H.-J., Schraft, R. D.
  Industrieroboter Kataloge von 1984 bis 1998
  Stuttgart, 1984-1998

- Warnecke, H.-J. Schraft, R. D.
  Industrieroboter. Handbuch für Industrie und Wissenschaft
  Heidelberg u. a. 1990